Thinking about Science, Reflecting on Art

Thinking about Science, Reflecting on Art: Bringing Aesthetics and Philosophy of Science Together is the first book to systematically examine the relationship between the philosophy of science and aesthetics. With contributions from leading figures from both fields, this edited collection engages with such questions as:

- Does representation function in the same way in science and in art?
- What important characteristics do scientific models share with literary fictions?
- What is the difference between interpretation in the sciences and in the arts?
- Can there be a science of aesthetics?
- In what ways can aesthetics and philosophy of science be integrated?

Aiming to develop the interconnections between the philosophy of science and the philosophy of art more broadly and more deeply than ever before, this volume not only explores scientific representation by comparison with fiction but extends the scope of interaction to include metaphysical and other questions around methodology in mainstream philosophy of science, including the aims of science, the characterisation of scientific understanding and the nature of observation, as well as drawing detailed comparisons between specific examples in both art and the sciences.

Otávio Bueno is Professor of Philosophy and Chair of the Philosophy Department at the University of Miami, USA.

George Darby is Postdoctoral Research Fellow at the University of Oxford, UK.

Steven French is Professor of Philosophy of Science at the University of Leeds, UK.

Dean Rickles is Professor of History and Philosophy of Modern Physics at the University of Sydney, Australia.

Thinking about Science, Reflecting on Art

Bringing Aesthetics and Philosophy of Science Together

Edited by Otávio Bueno, George Darby, Steven French and Dean Rickles

Routledge
Taylor & Francis Group
LONDON AND NEW YORK

First published 2018
by Routledge
2 Park Square, Milton Park, Abingdon, Oxon OX14 4RN

and by Routledge
711 Third Avenue, New York, NY 10017

Routledge is an imprint of the Taylor & Francis Group, an informa business

British Library Cataloguing-in-Publication Data
A catalogue record for this book is available from the British Library

Library of Congress Cataloging-in-Publication Data
A catalogue record for this book has been requested

ISBN: 978-1-138-68732-5 (hbk)
ISBN: 978-1-315-11492-7 (ebk)

Typeset in Bembo
by Deanta Global Publishing Services, Chennai, India

Contents

Notes on Contributors

Ann-Sophie Barwich holds a postdoctoral position as Presidential Scholar in Society and Neuroscience at the Center for Science and Society, Columbia University, having received her PhD at the University of Exeter. She works on olfaction as an emerging model system for neuroscience and the senses. Recent publications include 'Making Sense of Smell', *The Philosophers' Magazine* 73, 41–47, 2016 and 'Bending Molecules or Bending the Rules? The Application of Theoretical Models in Fragrance Chemistry', *Perspectives on Science* 23, 443–465, 2015.

Otávio Bueno is Professor of Philosophy and Chair of the Philosophy Department at the University of Miami. In 'Representing and Picturing: Approaches in the Sciences and the Arts' (*American Society for Aesthetics Newsletter*, 2014) and in 'How Theories Represent' (with Steven French; *British Journal for the Philosophy of Science*, 2011) he examines connections between representation in art and science. He has also written on the role of images in scientific reasoning ('When Physics and Biology Meet: The Nanoscale Case', *Studies in History and Philosophy of Biological and Biomedical Sciences*, 2011). He is editor-in-chief of *Synthese* and the Synthese Library book series.

Alix Cohen is a Chancellor's Fellow at the University of Edinburgh, having previously taught at the Universities of Leeds and York. She is the author of *Kant and the Human Sciences: Biology, Anthropology and History*, Palgrave, 2009, as well as numerous articles and book chapters on Kant, including 'Kant on the Possibility of Ugliness', British *Journal of Aesthetics*, 53(2), 199–209, 2013.

George Darby is a Postdoctoral Research Fellow at the University of Oxford. His research interests include the intersection of philosophy of science and metaphysics. In 'Representing Indeterminacy in Art and Science', *American Society for Aesthetics Newsletter*, 2014, he examines different ways in which indeterminacy occurs in art and on certain interpretations of quantum mechanics, and in 'Quantum Mechanics and Metaphysical Indeterminacy', *Australasian Journal of Philosophy*, 2010 and 'Vague Objects in Quantum

Mechanics' in K. Akiba and A. Abasnezhad (eds.) *Vague Objects and Vague Identity*, 2014, he examines issues in the formal representation of this kind of indeterminacy.

Catherine Z. Elgin is Professor of the Philosophy of Education at Harvard Graduate School of Education. She is the author of *True Enough, Considered Judgment, Between the Absolute and the Arbitrary, With Reference to Reference*, and co-author with Nelson Goodman of *Reconceptions in Philosophy and Other Arts and Sciences*.

Steven French is Professor of the Philosophy of Science at the University of Leeds and Co-Editor-in-Chief of *The British Journal for the Philosophy of Science*. His most recent publications include 'There Are No Such Things as Ordinary Objects' in J. Cumpa and B. Brewer (eds.), *The Nature of Ordinary Objects*, Cambridge University Press, forthcoming and *The Structure of the World: Metaphysics and Representation*, Oxford University Press, 2014 (pbk 2016).

Roman Frigg is Professor of Philosophy at the London School of Economics and is Director of the Centre for Philosophy of Natural and Social Science (CPNSS) as well as Co-Director of the Centre for the Analysis of Time Series (CATS) at the LSE. He has published widely on issues in the philosophy of physics and the philosophy of science. His recent work includes 'Models and Representation' with James Nguyen, in Magnani, Lorenzo and Bertolotti, Tommaso (eds.), *Springer Handbook of Model-Based Science*, Springer International Publishing, 2016 and 'Philosophy of Climate Science Parts I and II: Modelling Climate Change' with Charlotte Werndl and Erica Thompson, *Philosophy Compass*, 10(12), 953–964, 965–977, 2016.

James Nguyen is a Postdoctoral Researcher in Philosophy of Science at the John J. Reilly Center at the University of Notre Dame and received his PhD from the London School of Economics. His recent publications include 'On the Pragmatic Equivalence between Representing Data and Phenomena', *Philosophy of Science*, 83(2), 171–191, 2016 and 'The Fiction View of Models Reloaded' with Roman Frigg, *The Monist*, 99(3), 225–242, 2016.

Martin Pickup is a Postdoctoral Fellow in The Metaphysics of Entanglement research project at the University of Oxford and Junior Research Fellow at New College. He works on topics in metaphysics, early modern philosophy and philosophy of religion. His recent papers include 'A Situationalist Solution to the Ship of Theseus Puzzle', *Erkenntnis* 81(5), 973–992, 2016, 'Unextended Complexes', *Thought* 5(3): 257–264, 2016 and 'The Trinity and Extended Simples', *Faith and Philosophy* 33(4), 414–440.

Dean Rickles is Professor of History and Philosophy of Modern Physics and Australian Research Council Future Fellow at the University of Sydney, where he is also Co-Director of the Centre for Time. He has written several books, including most recently *A Brief History of String Theory*, Springer, 2014 and *Philosophy of Physics*, Polity Press, 2016.

Jon Robson is a Teaching Associate at the University of Nottingham. His current research focuses primarily on the epistemology of aesthetic judgements and, in particular, on the status of testimony in aesthetics. He was previously a postdoctoral fellow on the project 'Method in Philosophical Aesthetics: the challenge from the sciences' and is Co-Editor of the resulting volume *Aesthetics and the Sciences of Mind*, Oxford University Press. His publications include 'Video Games as Self-Involving Interactive Fictions' in *The Journal of Aesthetics and Art Criticism* (co-authored with Aaron Meskin) and 'Religious Fictionalism and the Problem of Evil' in *Religious Studies*.

Julia Sánchez-Dorado is working on her PhD at University College London. Her research is focussed on the inter-relationships between philosophy of science and aesthetics, particularly concerning the problem of representation. Her most recent presentations include 'Judgments of Similarity in Modeling Practices', *Models and Simulations* 7, 2016 and 'Misrepresentation and Similarity. Philosophy of Science meets Aesthetics', one-day conference, *Representation in Science and Art*, University of Manchester, February 2016.

Adam Toon is Senior Lecturer at the University of Exeter, having previously been a Postdoctoral Fellow at the University of Bielefeld in Germany. His recent work includes *Models as Make-Believe: Imagination, Fiction and Scientific Representation*, Palgrave Macmillan, 2012 and 'Fictionalism and the Folk', *The Monist*, 99, 280–295, 2016.

Introduction

It goes without saying (but we'll say it anyway!) that the two central human endeavours represented by 'art' and 'science' appear to be very different in all sorts of obvious ways. Yet they also display some, perhaps less obvious, similarities and these differences and similarities have of course been extensively discussed and explored in philosophical literature and elsewhere. However, the connections between the *philosophy* of science and aesthetics have yet to be treated in any systematic way, if at all. This is unfortunate, since such connections have the potential to reinvigorate core questions in both areas, especially those concerning matters of methodology and relationships with other fields. We believe that this book is the first to examine these relationships in significant depth and using detailed case studies.

The nature of representation is clearly one of the more obvious overlapping themes on which fruitful dialogue can be expected and, indeed, there has already been significant and detailed discussion here, particularly over the last several years in the philosophy of science. Both certain kinds of artistic work and scientific theories have been taken to represent the world, or some possible world, in a particular way. Immediately, philosophical questions arise on both sides: how do scientific models relate to the real world? How does a work of fiction relate to the fictional world that it creates? How does a painting represent its subject? Relatedly, we can ask what is involved in interpreting a theory, or a work of fiction or, indeed, a piece of music. And there are yet further questions that can be posed: we talk of scientific research programmes, on the one hand, and 'schools' of or 'movements' in art – to what extent do these have similar features? Can we talk of revolutions in art in the same terms as we understand revolutions in science? And what is an artwork? In the case of paintings and sculptures the answer may seem obvious, but what about a novel, or a piece of music? Again, similar questions arise with regard to theories and models.

The similarities between all these different kinds of questions and the answers that have been put forward encourage the transfer of devices and manoeuvres from the philosophy of art into the philosophy of science and vice versa (although perhaps to a lesser extent). Thus, philosophers of science have drawn significantly on both certain kinds of moves and case studies that can be found in the philosophy of art, taking in well-known accounts of representation in the latter domain as well as even better known examples of artworks. The further but over-arching question immediately arises: to what extent are such moves, devices and examples from the world of art and the philosophy of art pertinent when it comes to the kinds of moves and examples analysed in the philosophy of science? To what extent do Newton's mechanics, for example, represent in the way that Constable's *The Haywain* (supposedly) does? To what extent are examples such as the latter useful when considering representation in science? And although much of the traffic has gone in one direction, particularly recently, not all of it is one-way. Some philosophers of art have adopted more formal accounts of representation that look quite familiar to philosophers of science, and again we may ask: to what extent is the importation into the world of art accounts of representation involving devices like 'isomorphism' legitimate?

And, of course, as the above questions suggest, we might broaden our scope and consider not just representational or depictive paintings but other forms of art. As already indicated, there has been considerable discussion, especially recently, of approaches to fictions in art, science and mathematics with, in particular, significant consideration of the way in which our understanding of scientific modelling might draw on approaches to fictions in the philosophy of literature. Again we might ask: to what extent can the kinds of idealisations we find in a model of the movement of a spring, say, be regarded as fictional in the way that Bilbo from *The Hobbit* is? Are the kinds of accounts of fictions that we find in the philosophy of art – such as, famously, Walton's 'pretence' theory – straightforwardly applicable to models and theories in science?

What should we then say about the ontological status of such fictions, whether artistic or scientific? The standard recourse to *abstracta* leaves those of nominalist inclinations cold, of course. But such ontological concerns also apply to theories and artworks themselves. Consider the further questions: What is a scientific theory? What is an artwork? And what is a musical work? Such questions form the bread and butter of researchers in their separate disciplinary domains, but the potential links between reflections on theories and works of art and music have not been widely pursued, if at all. In particular, can the idea that musical works should be regarded as 'abstract artefacts' – in the sense of *abstracta* that are created and sustained by our intentions (as Amie Thomasson defends) – be imported into the philosophy of science? Here one might draw an obvious connection with Karl Popper's idea that theories are like pieces of music or works of literature by virtue of inhabiting his 'World 3', while acknowledging that abstract artefacts do not have the kind of independent existence that 'World 3' entities appear to have. Again, there are clearly numerous other connections that can be explored.

However, a broad concern that one might have, and that might keep the two disciplines – philosophy of science and aesthetics – from deeper interaction, is that these questions may have, at the bottom, nothing essentially to do with either art or science. The questions are questions of semantics, or logic, or epistemology or metaphysics; thus, they are applications at opposite sides of the periphery of the 'core' philosophical themes. This leads to the suggestion that philosophers of art and philosophers of science should really each be talking (separately) to the metaphysicians or the epistemologists. In a sense, this is true, because every philosophical question involves these core areas – that is why they are the core areas! And building separate bridges between the philosophy of science and metaphysics, say, on the one hand, and between aesthetics and metaphysics on the other, may be a viable approach. But consider: first of all, this would fail to directly accommodate or explore the similarities and possible connections between the two fields. Of course, those similarities and further connections may be illuminated indirectly, via each discipline drawing on metaphysical devices and manoeuvres, say, in their own way and the results then compared. But there may be much to be gained by first identifying and exploring those points of comparison and then looking to metaphysics, or logic or whatever to resolve the issues that arise. And secondly, there is a bluntly pragmatic reason of intellectual labour saved by drawing on such devices in a joint enterprise, as it were, rather than separately. Again, take the example of the comparison between Popperian entities and abstract artefacts. On the basis of that comparison, both kinds of entities face similar issues with regard to their construction or 'discovery', their sustained existence, their relative dependence on intentions, beliefs or other mental states. Tackling such issues from a common base may not only help in their resolution, but might also then indicate where possible differences may lie.

Our intention here is to develop these interconnections between the philosophy of science and the philosophy of art both more broadly and more deeply, including but also extending beyond the comparison with representations, fictions and the like, and exploring, as already mentioned, metaphysical, epistemological and methodological comparisons. In particular, we hope to have avoided the criticism, recently levelled at the former kind of comparison, that such inter-relationships are only superficial by encouraging the articulation of these comparisons in the context of extensive and detailed case studies.

We begin, then, with Julia Sánchez-Dorado's essay, 'Methodological lessons for the integration of philosophy of science and aesthetics: The case of representation'. Here she kicks off our discussion with an overview of current attempts to draw parallels between art and science, focussing on representation. Noting the use of examples from art to motivate or criticise certain views regarding representation in science, she argues that such moves are often 'contingent, sporadic or even misleading'. Furthermore, she insists that if we don't take into consideration certain methodological issues regarding the nature and purpose of such moves, profound problems may arise. Thus, she notes that the example of Picasso's *Guernica* is deployed in certain accounts of scientific representation (those of Chakravartty, French and Suárez, in particular) but without paying due regard to the artistic context or, especially, the material decisions made by Picasso during the creation of the work – decisions that impact directly our understanding of the relevant representational relations. Similarly, certain concepts from art are appropriated and applied to representation in science without paying due regard to the context from whence they came. Thus, the degree of depiction in art is taken to be equivalent to the degree of approximate truth in science, but art, of course, does not take complete depiction to be a goal, unlike – in this account – science, as standardly understood when it comes to truth (although some may disagree). Moving up to the meta-level, certain methodological moves in the philosophy of art may be appropriated and deployed in the philosophy of science. Thus, Sánchez-Dorado notes, French draws on Budd's isomorphism-based account of depiction to claim that representation in both science and art can be accommodated within such an account. However, leaving aside the point that Budd and French are actually addressing different questions, the notions of isomorphism in play originate in different frameworks and are used to respond to very different – in fact, opposing – views. These critical points bear on the issue of what methodological lessons we can learn with regard to the integration of art and science. Here she draws on recent work in the integrated-history-and-philosophy-of-science movement to highlight her claim that it is only by acknowledging the differences between the fields that dialogue between them can be fruitful. And she concludes by outlining two recent attempts at such an integrative project by van Fraassen and Elgin. Both illustrate that an appropriate methodological strategy together with clear epistemic goals are required to underpin an interdisciplinary approach of this kind.

In her own contribution, 'Nature's handmaid, art', Elgin defends the claim that art, like science, 'embodies, conveys and often constitutes' understanding, in the sense that that both use the same symbolic resources for the same ends. Focussing on truth, she notes that science employs models, idealisations and thought experiments that do not purport to be true. And such devices should not be dismissed as mere heuristics, since they are effectively ineliminable from scientific practice. Hence, those practices diverge from literal truth, and if that's fine for science then it should be likewise for art. After all, we often come away from engaging with a work of art feeling that we have learned something. But then, she asks, what exactly does that involve? Her answer is that works of art may reorient us in the sense of enabling us to see things differently from the way we saw them before. In defending this claim, Elgin explores the similarities between art and science, particularly with regard to exemplification: Laboratory experiments exemplify features of the world that go beyond the lab; Haydn's 'Farewell' symphony exemplifies the instability and tempestuousness of our lives. And this exemplification is not just a matter of making manifest that which is already known; rather, unexpected and surprising features may be revealed. Likewise thought experiments, whether in science or in literature, exemplify certain features, and by distancing us from the phenomena enable us to see them more clearly. And this distancing may not be merely pragmatically useful but essential to the exemplification. It is by virtue of the artificial nature of practices, such as the Miller-Urey experiment that indicated how life was formed, that we learn about significant features of the world. And, finally, exemplification plays a crucial role in representation: by exemplifying certain features and ascribing them to what is being denoted, we obtain 'representation-as'. Thus, by exemplifying the capacity for a distinctive mode of oscillation and ascribing it to a spring, a certain diagram or model represents the spring as a harmonic oscillator. Nevertheless, there are differences, with regard to the precision of our representations and the agreement, or lack thereof, regarding interpretation. But these, Elgin insists, do not pertain to the question of whether the two disciplines advance understanding, but rather to how they do so.

Frigg and Nguyen also focus on representation in their essay, 'Of barrels and pipes: Representation-as in art and science', where they present both the movie *The Way Things Go* and the Philips-Newlyn model of economics as examples of representations-as, and then ask the question, what is the basis for this sort of representational relationship? In response, they generalise Goodman and Elgin's account of representation-as in art to yield what they call the DEKI account, which analyses representation-as in terms of denotation, exemplification, keying up and imputation. A model denotes a target system, and this establishes that the model is a representation-of that system. But this is not sufficient for representation-as. Here the core idea is that the properties exemplified in the model are then imputed to the target system. But the latter are typically not exactly the same as the former, and so we need a key that specifies how they are related. This account can then be applied not only to scientific models like the Philips-Newlyn machine but also to symbolic art, as in the case of Frans Pourbus the Younger's painting of Anne of Austria. This is both a Princess-with-dog-representation and a representation-of Princess Anne, because it denotes the princess. But, on their view, it is not a representation-of her dog, as the part of the painting showing a dog does not denote anything. Nevertheless, the dog is an important part of the picture and falls under 'exemplification', with the relevant 'key' being the convention that associates a dog with fidelity. The painting then imputes the thus keyed-up property to the princess and represents her as faithful. Furthermore, this framework applies to fictional or 'imagined' objects, whether scientific models or artworks such as works of literature. Finally, Frigg and Nguyen conclude by noting some of the differences between artistic and scientific representations, although they insist these are more of degree than of kind.

Pursuing the representational theme, in her paper, 'Is Captain Kirk a natural blonde? Do X-ray crystallographers dream of electron clouds? Comparing model-based inferences in science with fiction theory', Barwich argues that, like fictions, scientific models have an ambiguous representational relationship to the world, in that it may just not be clear whether a given feature of the model represents some feature in the world or is an artefact of the model. Yet, compared to fictions, we typically assign a special epistemic role to scientific representations. Her detailed case study of the way that protein models are constructed allows Barwich to examine the 'fictionalisation strategies' used by scientists and to illuminate their limitations. Here, she claims, a form of pluralism that involves the integration of the model and associated strategies, such as idealisation, with other methods and procedures, such as material manipulations, helps to separate the representational from the fictional elements. The overall conclusion is that the function and limitations of scientific representation must be examined from such a pluralistic perspective, in the relevant context.

Similar kinds of issues dealing with what it is that we take our theories to be about are explored in Bueno's paper, 'Interpreting the Sciences, Interpreting the Arts', in which, as the title suggests, he focuses on the notion of *interpretation*. When it comes to science, he contrasts two different views of interpretation: the model-theoretic, according to which an interpretation is simply an assignment of truth-values to the sentences of a theory, so that all of its sentences come out true, as well as the 'modal', according to which an interpretation indicates the way the world could be if a given theory were true. The former suffers from certain inherent representational limitations when compared with the latter. Likewise, in the arts, we can distinguish between a 'metaphysical' account of interpretation, which takes an interpretation to be whatever is ultimately required to make the claims (implicitly or explicitly) made by an artwork come out true, and a 'making sense' view, according to which an interpretation provides an account that ultimately makes sense of the artworks in question. Drawing on a range of examples, from Gabriel García Márquez's *One Hundred Years of Solitude* to Rembrandt's self-portraits and Mark Tansey's painting *The Innocent Eye Test*, Bueno compares these accounts and concludes that the former account is also deficient in that it cannot accommodate artworks that might be understood as ironic, for example. The latter account, on the other hand, is not so constrained but might be seen as too open-ended; hence, like Barwich, Bueno recommends a form of pluralism, which also mitigates certain worries about the ontological commitments of such interpretations, particularly when it comes to fictional entities. Focussing now on the example of films, Bueno emphasises the crucial role of interpretation and, in particular, a certain kind of imagination in making sense of the visual content of a movie such as *Blade Runner*, for example. Bringing the discussion back round to the sciences, he concludes that a similar kind of perceptual imagining also plays an important role in the interpretation of scientific results, as in the use of electron microscopes and other devices.

The nature of truth in fictional works, whether literature or film, is also discussed in Darby, Pickup and Robson's essay, 'Deep Indeterminacy in Physics and Fiction'. There they identify a comparatively seldom-discussed form of indeterminacy in fiction, in response to common questions in the philosophy of art, such as: is the governess in James's *The Turn of the Screw* haunted by supernatural apparitions or merely by symptoms of her own mental instability? Their answer posits a form of 'deep' indeterminacy, springing from the structure of the artwork itself. In this respect, Darby, Pickup and Robson argue, it is akin to the kind of indeterminacy that arises in quantum physics. In the latter, it is cashed out via certain mathematical constraints on possibility; in the former, the indeterminacy is formulated by the appeal to the relevant identity constraints imposed on works of fiction in the sense that to be that very story is to contain indeterminacy of this sort. They further argue that this kind of deep indeterminacy cannot be captured by the standard accounts in modal metaphysics, since such accounts rely on explicating the indeterminacy via complete, possible worlds. Instead they suggest that it should be modelled via 'situations', understood as parts of possible worlds, and their partial nature then nicely accommodates the lack of truth value of certain propositions. Replacing possible worlds with situations in the extant frameworks for modelling indeterminacy then allows us to capture this deep form in both the case of quantum physics and fictional artworks. And, thus, we have a nice example of mainstream analytic metaphysics providing the kinds of tools that can then be naturalistically applied across both art and science.

Rickles also considers the application of certain tools, or the trade of certain good approaches from one philosophical domain to another, in his 'Some Philosophical Problems of Music Theory (and some Music-Theoretic Problems for Philosophy)'. Here he explores certain avenues along which the philosophy of science may be applied to the philosophy of art, specifically the philosophy of music. In particular, he considers the issue of the nature of musical laws and theories and asks whether the kinds of analyses undertaken by philosophers of science when it comes to scientific theories and models might also be applicable here. Focussing primarily on metre and tonality, he presents an overview of some of the highly mathematical treatments of these features, noting the role of certain invariances, which of course fall under the general notion of symmetry, which in turn plays a crucial and prominent role in modern physics. Thus, on the one hand, in this respect theories of music lie close to those of science and might be expected to be equally worthy of treatment by philosophers of science. However, on the other, metre and tonality are also highly subjective, and hence the question arises as to what the relevant theories and models are *of* – some objective structure or subjective experience? What then is 'musical reality', and what is musical theory *about*? Here Rickles draws on a further and perhaps rather unexpected bridge with the foundations of space-time theory, where we find a similar kind of 'bootstrapping' between matter and space-time as can be identified as holding between the 'rhythmic surface' and metre. Such a bridge then allows him to export into the philosophy of music a particular approach developed in response to modern space-time physics, namely Eddington's 'subjective structuralism'. On this view, matter and space-time – and, by analogy, metre and rhythmic surface – come as a package, as the mind selects certain structural features of the world (and here there is an obvious comparison with the Kantian perspective, although Eddington himself was keen to distinguish his view from Kant's). Likewise, metre is both part of the essential structure of a piece of music, yet we do not directly hear it but essentially construct it. Thus, our theories of music are about this 'package' of the subjective and objective rather than some unadulterated and unobtainable musical 'reality'. As Rickles says in his conclusion, whatever one thinks about this Eddingtonian stance, it must be acknowledged that there are many fruitful interconnections to be explored between the philosophy of music and the philosophy of science.

Pursuing the Kantian theme, in her 'Kant on Beauty and Cognition', Cohen examines Kant's twin claims that, first, cognition cannot be beautiful since it has to do with concepts while beauty is inherently non-conceptual, and, second, beauty cannot contribute to cognition since it is based on subjective feelings while that latter is all about objective knowledge. On the contrary, she argues, cognition can be beautiful, and beauty is cognitively valuable, because the experience of beauty stimulates our cognitive powers and thereby enhances our cognitive activity. Note, however, that this does not mean that the beauty of a cognition provides some kind of epistemic guidance or allows us to choose between different theories; rather, beauty contributes to cognition indirectly, via its effect on the activity itself rather than the content of cognition. Nevertheless, she insists that beauty has an epistemic function in that it acts as an indicator of cognitive efficacy by virtue of enhancing cognitive activity – a conclusion that obviously resonates with certain views adopted in current discussions of the role of aesthetic factors in science.

Such activity is obviously the subject of a specific scientific discipline, namely cognitive science, and Toon, in his 'Epistemology as Fiction', examines how we might respond to the challenge of accounting for the role of folk psychology in this context. One option is to adopt Churchland's eliminativism, but that is of course a hard row to hoe. Alternatively, we might again adopt a fictionalist line and regard the relevant folk psychological states as useful fictions. Moving in the other direction now, from philosophy of art to the philosophy of science, and drawing on Walton's analysis of art and fiction, Toon argues that such talk about states can be understood to involve pretence within a game of make-believe. This then offers an advantage over eliminativism in that it allows us to retain our ordinary language for talking about the epistemic activities of scientists, say, without requiring that such talk be vindicated by whatever turns out to be the final theory of cognitive science. Here we see how an approach towards fiction in art may be appropriate and deployed in a much broader epistemological context.

Concluding the volume, French's 'Art, science and abstract artefacts' pursues the idea of trading across domains via his consideration of the question: what *is* a scientific theory? There is a current stream of thought within the philosophy of science that takes theories and models to be abstract entities, like 'never constructed buildings described in architectural drawings'. Relatedly, as already noted, Popper famously assigned theories to his 'World 3', distinct from the worlds of both material objects and mental states. Here he also situated certain artworks, such as works of literature (e.g. Hamlet) and music (Beethoven's Fifth), crucially both as capable of change and as causally interactive entities. Such a view bears comparison with Thomasson's more recent abstract artefacts account, according to which, as noted, certain artistic objects, such as works of music and literature, lack a spatial location but can be regarded as temporal objects in that they are still created, come into existence, change and may cease to exist. However, French raises a number of concerns: how exactly do the kinds of creative practices we can track in both art and science generate such entities? And which practices? Over what time frames? Appeal to the relevant intentions involved runs afoul, on the science side, of the apparent phenomenon of multiple discovery. Although this does not quite draw as clear a distinction between artworks and theories as might first be thought, French argues that theories exhibit greater 'modal flexibility' than artworks with regard to their creation or discovery, but this further impacts on their status as abstract artefacts. In the end, he suggests we can cut through this knot of concerns and issues by simply denying that theories are entities to begin with, drawing on Cameron's eliminativist stance towards music works.

There is much more to say, of course, about all of these themes and ideas as well as about the essays themselves. But we hope that this collection will give some indication of just how fruitful such explorations of the multiple interconnections between the philosophy of art and of science can be. And perhaps it will encourage others to not only take such explorations further, but also open up new avenues of investigation and new modes of transfer between the philosophy of art and the philosophy of science.

1 Methodological lessons for the integration of philosophy of science and aesthetics

The case of representation

Julia Sánchez-Dorado

1 Introduction

Philosophers of science and philosophers of art have increasingly joined forces to tackle a range of contemporary debates in epistemology and metaphysics. This expanding literature is in fact a particular manifestation of a much broader trend that has tried to bring art and science together in modern times, not only through a philosophical lens but also by tracking shared historical episodes and identifying common practices. In this paper, I will focus on the analysis of a specific subset of that heterogeneous literature that tries to connect art and science, namely, recent attempts to integrate discussions on scientific and artistic representation.

Particularly in contemporary philosophy of science, there is a manifest tendency to refer to examples and concepts from art to illustrate or support arguments concerning scientific representation. In 2006, Callender & Cohen wrote a paper entitled 'There is no special problem about scientific representation', where they openly questioned the genuineness of the problem of *scientific* representation and urged philosophers to examine it in the context of wider debates in philosophy of language, philosophy of mind or aesthetics. Other attempts to connect the problems of scientific and artistic representation can be found in Suárez (1999, 2003, 2004), French (2003), van Fraassen (2008), Downes (2009), Elgin (2010, 2011, 2017), Chakravartty (2010) and Ambrosio (2013). In addition, there is a flourishing literature on fictionalism that has tried to incorporate modern theories of literary fiction into accounts of representation in science.[1]

Some reasons can be suggested to explain the increasing interest in including elements from art in the debate of scientific representation. One of them relates to the emergence of studies of models in contemporary philosophy of science. The shift towards the study of models, moving away from the idea that scientific theories were the most fundamental units of science, began in the 1960s and gained momentum in the 1980s and 1990s. Philosophers of science usually agree that, despite the great variety of models there are – material, graphical, mathematical – something they have in common is that they are not linguistic entities, at least not in the same way scientific theories were taken to be in the syntactic view. Thus, by emphasizing the non-linguistic character of models, commonalities with artistic products such as paintings, photographs or sculptures emerged. Moreover, these commonalities became particularly useful when philosophers of science investigated the nature of particular manifestations of models such as scientific images, diagrams, scale models or computer simulations.

Another motivation for the reference to pictorial arts in accounts of scientific representation seems to be the more general acknowledgment of the limitations that philosophy of science has when addressing particularly complex problems like representation. Recognizing the strengths of other traditions of thought in dealing with analogous problems can be taken as an act of academic humility, linked to a positive view of what interdisciplinary work can do for contemporary research.

However, the potential benefits of integrating aesthetics and philosophy of science to address the problem of representation are diminished when references to art in this literature happen to be too contingent, sporadic or even misleading. Connections between science and art are occasionally advocated, but without openly questioning how they are justified in epistemological terms and what the gain of doing so is. The field of aesthetics has a long tradition of asking questions like: how do pictures represent? How do photographs stand for particular targets in the world? What is the role of similarity in depiction? That tradition has to be carefully and systematically considered if the aim is to establish fruitful links with debates on scientific representation. One of the goals of this paper is to show some problematic consequences of not taking explicitly into consideration important methodological issues concerning how and to what purpose elements from art are incorporated into the debate of scientific representation. The other goal is to vindicate the potential epistemological benefit of bringing aesthetics and philosophy of science together, once the previous methodological issues are adequately considered.

In Section 2 of the paper, I identify three different ways of incorporating elements from art into contemporary debates of scientific representation and point out some of the limitations they respectively present. Recent papers by Suárez (2003, 2004), Chakravartty (2010) and French (2003) will help illustrate this point. In Section 3, I present some methodological reflections about the integration of philosophy of science and aesthetics, and interdisciplinary work more broadly. Two accounts in recent literature will be discussed to show how the integration of debates in philosophy of science and art can be particularly fruitful to address the problem of representation, namely van Fraassen's (2008) and Elgin's (2010, 2011, 2017) latest proposals.

2 Three constraints on the relation between scientific and artistic representation

It is possible to observe philosophers of science including elements from art in their accounts of scientific representation in at least three different ways.

Each of these ways can help bring into light important commonalities between scientific and artistic products, but can also convey some methodological difficulties. The first one is the use of artworks to illustrate certain features of scientific representations. The difficulty in this case can arise when there is no explicit consideration to the aesthetical and historical background in which those artworks were produced. The second one is the use of concepts from theories of modern art to make claims about representation in science. The difficulty here appears when the original meaning of those concepts is partially misleading. And the third one is the establishment of a direct link between accounts of representation in aesthetics and philosophy of science. In this case, the potential of the strategy can be substantially diminished if there is not enough consideration of the underlying worries that philosophers in each field have.

The first situation refers to the occasional allusion to particular artworks to highlight or uncover a feature of scientific representations. An example of this can be found in the recurring references to Picasso's *Guernica* in recent literature on philosophy of science (see Suárez, 1999, 2003, 2004; French, 2003; Chakravartty, 2010). As Ambrosio (2013) has claimed: 'these cursory references to Picasso's painting occasionally appear in philosophy journals to support, in strangely instrumental ways, entirely contrasting approaches to scientific representation' (Ambrosio, 2013: 109). Suárez (2003, 2004) uses the example of *Guernica* to defend a deflationary, inferential conception of representation. French (2003) refers to Picasso's painting to support a model-theoretic account of representation based on partial isomorphism. Chakravartty (2010) describes the representational relations in *Guernica* as sustaining an 'approximate truth' conception of the goals of science and art. How is it plausible that *Guernica* strengthens each of these accounts of representation?

Following Ambrosio (2013: 109), we should probably admit that none of these accounts does complete justice to the representational relations governing *Guernica*. References to the painting appear rather in isolation in these works, maybe accompanied by some historical information around the piece, but not taking into consideration some of its defining aspects. Among those aspects, we should include the material decisions made by Picasso during the creation of the work, the specific place of the painting in the history of art (for instance, how it follows the principles of Cubism and how Cubism fits in the panorama of the Avant-gardes) and central aesthetic issues related to the piece more or less directly, such as what abstraction meant for traditional styles of depiction. The status of a painting as an artwork and, more importantly here, as a *representational vehicle* cannot be accurately discussed without considering these central aspects. Fortunately, in the case of *Guernica*, the painting is well documented and has been extensively studied by historians of art (see Arnheim, 1962; Chipp, 1988; Oppler, 1988). Ambrosio specifically suggests that a closer inspection of the process culminating in the final version of *Guernica* – through its over 40 preparatory sketches – can reveal a different and far more interesting story about the practices of representing that underpin it (Ambrosio, 2013: 109).

To give a more specific example, Suárez (2003) uses the case of *Guernica* in the context of defending a specific claim about scientific representations: that similarity is not a constituent of the relation of representation (Suárez, 2003: 233–6). His main concern (Suárez, 2003) is with approaches in philosophy of science that take either similarity or isomorphism to be necessary and/or sufficient conditions for representation. And it is in the pictorial arts where he finds particularly suitable examples to argue against these approaches. For instance, he shows that there are probably some figurative elements in *Guernica* that look like, or are similar to, certain things in the world, such as a weeping woman, a bull or a horse, that we would recognize in the composition. However, we would hardly say that those elements literally resemble, or are similar to, 'the threat of fascism', which is the much more intangible target of the painting. This fact demonstrates for Suárez that similarity is not necessary for representation, and more generally that the relation of representation cannot be explained by, or reduced to, a relation of similarity (Suárez, 2003: 236).

Even if the former analysis is undeniably correct, I argue that it is only a small part of the story that explains the representational relations governing *Guernica*. A more complete story would have not permitted the claim that similarity is not necessary for representation without further characterization of how, nevertheless, similarities of appearance are present and play a role in pictorial representations like this one. And this more complete story can only be given by taking into consideration the preparatory sketches of the painting and the aesthetic discussions surrounding its creation and reception, instead of only the piece as an analysable final product. Thanks to the historical record of successive sketches, we know that the aim of the painting was not realistic or figurative representation, as, for instance, Picasso decided to transform a quite realistic drawing of a horse in an early sketch into a more geometricized and abstract version of it (Ambrosio, 2013: 112). At the same time – and this is the key – we also observe that an important means of the continual practice of depicting was the search for relevant similarities that allow the composition to access its more intangible target: the rise of fascism as well as a universal statement against war. This is how the painting progressively gains in pictorial detail of some of its components, such as the mouths and the eyes of the characters depicted in the scene (the woman, the horse, the bull), to make them exaggeratedly and selectively resemble the expressions of people in great pain (Ambrosio, 2013: 112–14).

At this point, Suárez would claim that the previous issues concern the *means* of the representation, and not the *constituents*, which was his main worry when criticizing the role of similarity (Suárez, 2003: 230).[2] Nonetheless, he specifically stated that none of the visual similarities present in *Guernica* 'are a good guide to the actual targets of the representation' (Suárez, 2003: 236).

This is exactly what a more complete description of the painting contradicts: selective similarities might not constitute the relation of representation, but they are precisely 'good guides' to successfully access the target of the representation, which is anyway far too complex to be exhausted by similarity conceived as a point-to-point correspondence. In short, using cases from pictorial arts in an argument about the general conditions for scientific representation might not be in every case as effective as it might initially have seemed. Introducing an artwork as example, especially if we consider the story of the practices that culminate in it, could result in the challenge of some of the general claims at stake instead of in their plain validation. In this case, paying attention to the different components of *Guernica* might have the effect of triggering new questions about the relations between the *means* and the *constituents* of a representation, or about the goals that define the *means* used in practice.

The second way of introducing elements from art in the debate on scientific representation refers to the appropriation of concepts from art to make claims about representation in science. As part of the volume *Beyond Mimesis and Convention. Representation in Art and Science* (Frigg and Hunter, 2010), Anjan Chakravartty (2010) writes a paper entitled 'Truth and representation in science: Two inspirations from art'. There he uses terminology originating from theories of modern art and also incorporates examples from abstract and performance arts (including *Guernica* [2010: 45]) as heuristic tools to argue for an approximate truth conception of representation. 'I believe that analogies to practices of representation in art can serve as valuable heuristics towards understanding how and in what manner scientific representations can be true' (2010: 33). Chakravartty's initial intuition when he claimed this must have been that what scientific models and artworks have in common is that they depart in important aspects from the targets in the world they represent, and yet they succeed in providing true characterizations of those targets. Unfortunately, this intuition is never clearly articulated in the paper, and it turns out to be somewhat obscure at highlighting commonalities between scientific and artistic representations.

Chakravartty proposes a triple analogy: 'realistic' and 'non-realistic' styles of representation (Chakravartty, 2010: 40–41) are considered respectively akin to 'depiction' and 'denotation' in the arts, and to 'truth' and 'reference' in the sciences (2010: 45). This implies that there is a correspondence between realistic styles, depiction in art and truth in science. Since Chakravartty sees these notions as degrees in a spectrum, the closer to realistic styles we are, the closer to perfect depiction and truth, and the further from non-realistic styles, denotation and reference. But why assimilate degrees of depiction in art with approximation to truth in science? That would only make complete sense if the main goal of art was perfect depiction, since Chakravartty assumes that truth is the main goal of science. But this is clearly not the case for art. Otherwise, any pictorial style that is intentionally departing from figurative modes of depiction has to be understood as a failure in principle. Even if we accept that approximating truth is the goal of science, having a look to the history of modern arts shows that perfect depiction is not the goal of art.

That being said, it is hard to believe that Chakravartty is trying to claim exactly that about depiction, although the analogy invites us to think in this direction. Probably the reasoning behind the analogy is that Chakravartty considers that seeking or approximating truth is the goal of both science and art, as he affirms: 'truth in both domains should be understood in terms of approximating reality by means of representation' (2010: 34). Then, 'depiction' and 'denotation' would be two possible ways of approximating truth in the arts, while 'truth' (in the sense of true characterization of the target) and 'reference' would be two ways of approximating truth in science. Even in these terms, the analogy that results is ambiguous. For Chakravartty, the greater the number of relevant properties of a target system a representation describes (regardless of abstractions) and the more accurately it describes them (regardless of idealizations), the closer to truth that representation is (2010: 45). If this is right, idealizations and abstractions are understood as limiting or constraining aspects of representations that should be corrected or improved upon to approximate truth in science. This claim has had several detractors in contemporary philosophy of science (see Elgin, 2004; Elliott-Graves and Weisberg, 2014). At least some idealizations and abstractions in scientific models are not meant to be corrected by incorporating more properties, nor by yielding more information about those properties. It seems that the analogy with art Chakravartty uses could be supporting the opposite view to the one he is defending: the parallel with art reinforces the idea that abstractions and idealizations can be responsible for the success of representations, instead of being elements that need to be improved upon to reach higher levels of approximation to truth. *Guernica* would not be an *improved* representation of its target by pictorially describing more properties of its target, nor by figuratively yielding more information about those specific properties.

Furthermore, the way in which Chakravartty defends 'approximate truth' as the common goal of science and art, alluding to Nelson Goodman's *Languages of Art* (1968), can be challenged too. *Languages of Art* is probably the work in aesthetics most quoted by contemporary philosophers of science working on the topic of representation. It is mostly cited for its logical argument against similarity[3], while other important contributions in it, such as those concerning the common aims of science and art, are frequently overlooked. Chakravartty does bring some of these matters to the fore of the debate, but it is open to question whether the concept of approximate truth that he ascribes to Goodman is actually equivalent to the one he is defending.

At one point in *Languages of Art*, Goodman uses the phrasing 'arriving at the nearest approximation to truth' (Goodman, 1968: 263) and Chakravartty claims to see a parallel with his own idea (Chakravartty, 2010: 40). However, Goodman made quite clear that truth was neither the ultimate goal of art nor of science and that this was a crucial commonality between the two domains. Particularly in science, he argued, 'despite rife doctrine, truth by itself matters very little' (Goodman, 1968: 263). We have to exercise our judgment not on grounds of truth but on the basis of the simplicity or strength of the hypotheses we have. And *this* is the closest to an idea of 'approximation to truth' we can appeal to (Goodman, 1968: 263). In a quite different spirit, the concept of approximate truth that Chakravartty proposes is an objective one that can be measured on grounds of the number of properties that a representation shares with its target and the degree of accuracy of those shared properties (Chakravartty, 2010: 45). In conclusion, the attempt to highlight connections between scientific and artistic representations in the case of Chakravartty's paper is not as acutely successful as it might, had the concepts of 'realistic style', 'depiction' and 'approximation to truth' (in Goodman's sense) been as precisely problematized as they have originally been in aesthetics.

A third possible way of introducing elements from art in the debate on scientific representation is the combination of specific philosophical accounts in aesthetics and philosophy of science. The explanatory potential this strategy has can be reduced if the different underlying concerns that philosophers in each of the fields have are not explicitly considered. To illustrate this point, I refer to Steven French's (2003) paper 'A Model-Theoretic Account of Representation (or, I Don't Know Much About Art... But I Know It Involves Isomorphism)', where he draws on Malcolm Budd's (1993) thesis on isomorphism in artistic depiction. In doing so, French is engaging with an account in contemporary aesthetics that in principle embraces a very similar challenge to his own: explaining representation in virtue of a relation of structural similarity or isomorphism.

However, Budd's concerns diverge from French's in very specific but key aspects that make the analogy between the two accounts rather complicated. French's (2003) main focus of attention is with the constitutional question about representation, namely whether something is a scientific representation of something else. And an adequate answer to this question, he argues, could be given in terms of an isomorphism relation between the target and the source of a representation (French, 2003: 1473). In other words, isomorphism is a good candidate to be a necessary – and even sufficient – condition for representation. In a different vein, the debate in aesthetics in which Malcolm Budd (1993) is immersed assumes, almost without exception among its members, that the notion of similarity (also resemblance, isomorphism) has to be discussed in relation to a different question: 'how do representations depict?' or 'what distinguishes pictorial from non-pictorial representations?' (Budd, 1993: 217). Budd's account of isomorphism is not answering the question of what the constituents of representation are, but what makes representations pictorially accurate.[4] Consequently, the first key difference between the two accounts is that Budd is concerned with pictorial content while French is trying to define conditions for representation.

Now, for the sake of the argument let's assume that both Budd and French were trying to answer the same – and less problematic – question about the *means* of representation with their accounts of isomorphism. Even in that case, the notions of isomorphism they are invoking are originated in different frameworks and respond to dissimilar opponent views in their fields. In his 1993 paper, Budd starts by claiming that the 'obvious and familiar' idea that a picture represent its subject by means of the properties it shares with it is clearly 'naïve and inadequate as it stands' (Budd, 1993: 217). He rejects the old intuition of traditional resemblance theories of pictorial representation that 'for a picture to accurately depict a scene is for the picture itself to resemble the scene' (Greenberg, 2011: 47). In contemporary debates in aesthetics, more refined resemblance theories of depiction can be divided up according to two general conceptions, one that assumes that the similarity in question is real, and the other one that assumes that it is merely experienced (see Greenberg, 2011: 47n; Abell, 2009: 188).[5] The strategy that Budd as well as Peacocke (1987) and Hopkins (1998) adopt corresponds to the second conception. The talk on isomorphism in these approaches only makes sense in terms of internalized perception of the subjects: 'The appearance an object presents to the viewer is dependent upon the point of view from which it is seen, the manner in which it is illuminated, and the mental and visual apparatus of the viewer' (Budd, 1993: 233). Particularly for Budd, experiences of resemblance occur between the *design* of a picture and the *visual field* of the viewer (Budd, 1993: 221).[6]

The ambiguous element in French's argument is that he seems much closer to the 'objective similarity' accounts of depiction than to the 'experienced similarity' ones that Budd holds. French's notion of partial isomorphism invokes an idea of mapping that exclusively concerns the two objects of the representation, vehicle and target, one being structurally similar to the other. The role of the subjects in this equation is limited to the identification of the existing shared structures, and to fill in, if possible, the partial elements of the mapping, for instance, the relationships that have not yet been established to hold (French, 2003: 1481). He makes clear his 'objective' take on similarity when he gives the example of a supposedly natural representation of the Lorentz transformations in the sand, made by the sea and wind. Given the existent isomorphism between the marks carved in the sand and the physical system, French would unproblematically claim that 'the theory is there' (French, 2003: 1474). That is, the (partial) sharing of structures of a vehicle and a target is all that is needed to have a representation.[7] In sum, the central role Budd concedes to the subjects intervening in the process of representing reconfigures the notion of isomorphism, to the point that it means something considerably different to what French is assuming it to mean in this view.

Additionally, Stephen Downes (2009) noticed a further difficulty with French's account: relying on Budd's proposal was not doing any good for his view, since experienced accounts of similarity have been strongly criticised in recent years in aesthetics (Downes, 2009: 424). The reason for this criticism is the so-called problem of 'illegitimate internalizing of content' (Lopes, 1996: 23). Subjective similarity accounts postulate inner pictures with properties of the visual experience that they aim to explain. So these accounts fail to pass the independence challenge (the challenge of explaining how we experience pictures as like their subjects independently of knowledge of what they depict)

(Downes, 2009: 424; following Lopes, 1996: 23). The conclusion Downes extracts from this situation goes as follows: 'a pessimistic outlook is that lessons from aesthetics indicate that philosophers of science should look elsewhere for an account of representation for models' (Downes, 2009: 420). I believe we do not need to accept that pessimistic conclusion about the project of bringing philosophy of science and aesthetics together to dialogue about representation. But maybe one lesson to be learnt is that conceptual differences and internal debates carried out over years in specific philosophical disciplines should not be disregarded when comparing accounts in different fields.

By referring to different papers in recent literature as examples, I have described some ways by which philosophers of science introduce elements from art into their accounts of scientific representation, and some possible difficulties this can entail. More than merely pointing to shortcomings of these accounts, the aim was to bring attention to the important epistemological benefit that current attempts of dialogue between philosophy of science and aesthetics can bring, particularly when the dialogue is established in a thorough and systematic way. In their modern form, aesthetics and philosophy of science have existed as independent fields for at least one century, mainly as a consequence of the specialization of academic disciplines which was consolidated in universities at the dawn of the twentieth century.[8] So even if they are two branches within philosophy, philosophy of science and aesthetics are equipped with distinctive disciplinary tools that organize their agendas of discussion and their internal debates. Those distinctive tools should be explicitly considered, also in cases where the goal is to challenge the alleged differences between scientific, artistic and other kinds of representations. In the next section of the paper, I present some general methodological reflections about the integration of philosophy of science and aesthetics, and interdisciplinary work more broadly.

3 Methodological lessons for the integration

One way of interpreting the previous difficulties of including examples or theories from art in accounts in philosophy of science is by pointing out a common methodological issue, or more precisely, a general lack of explicit methodological debate about how and to what purpose discussions about scientific and artistic representation can be integrated. I mentioned at the beginning various possible motivations of contemporary philosophers of science to be interested in engaging with debates in aesthetics. But there are, beyond those motivations, some central questions of methodological character that have to be more openly discussed to make the integration of aesthetics and philosophy of science completely fruitful. For instance, 'what is the precise role that elements from art are playing in an argument about scientific models?' and 'what is the epistemological benefit of combining discussions on scientific and artistic representations?' We can respond to the first question affirming, for example, that elements from art are mere anecdotes within an argument about scientific models, or that they are particular examples to show parallels between scientific and non-scientific representations, or crucial elements to argue for a general theory of representation that equally describes all types of representations, etc. Making these remarks explicit can avoid ambiguities about the purpose of the integration and function as a guide to reading philosophers' arguments correctly.

The claim about the lack of explicit methodological discussion surrounding the inclusion of art in accounts in philosophy of science can be taken as part of the broader observation that methodological debates in philosophy of science are scarce. Phyllis Illari (manuscript) has observed, after attempting a systematic search for literature on methodological questions in contemporary philosophy of science, that there are very few debates of this kind in peer-reviewed journals, and only occasional comments in books and collections (Illari, manuscript: 3–4). She is particularly concerned with the argumentative strategies that philosophers of science adopt when they introduce case studies in their accounts: although vigorous methodological preferences are held, the justifications for their preferences are not openly debated.[9] Her paper is then a call for explicit debate, since 'confusion can be eliminated by greater methodological clarity' (manuscript: 23). Specifically her 'strongest methodological recommendation is that authors more explicitly articulate their aims, including how they choose and use their examples and cases in arguing for their thesis' (manuscript: 2).

Illari also noticed that the sub-field of iHPS (Integrated History and Philosophy of Science) is one of the places where most of the existing methodological debates in philosophy of science have occurred (Illari, manuscript: 6). Work in iHPS initiated in the 1960s and was profusely developed in the following decades.[10] This sub-field draws on the premise that interdisciplinary work is necessary for the advancement of more adequate descriptions of the scientific enterprise. However, integrating two different traditions like history and philosophy raised important methodological issues that needed to be openly addressed. The main issue was, of course, how to gather historical evidence about particular circumstances to justify philosophical claims, which have a universal (and even ahistorical[11]) character; or, vice versa, how to sustain general philosophical claims with reference to singular historical events. The point I would like to stress here, although it would need further characterisation in future work, is that in the last fifty years an open debate about the best methodological approach for iHPS has run in parallel to actual proposals in iHPS. Similarly, running methodological discussions about the integration of philosophy of science and aesthetics could have a positive impact on the endeavour of advancing particular accounts of scientific and artistic representation.

An example of the kind of methodological reflection in iHPS that could be illuminating here is found in Peter Dear's (2012) paper 'Philosophy of science and its historical reconstructions'. Dear (2012: 68) argues that at first sight we could judge iHPS as having a 'single subject matter', namely *science*, which is observed from two perspectives, one historical and one philosophical. If this was strictly the case, the integration of the two fields would happen in a not very problematic way, by combining descriptions from one discipline and the other. Contrary to this, Dear points out that different disciplines have their own problems and their own ways of addressing those problems: 'it is surely not clear that history and philosophy had the same way of doing things; that they saw their common subject matter, science, in the same way, or raised the same questions about it' (Dear, 2012: 68). His own proposal in iHPS is consequently directed to identify, in the first place, the particular underlying questions that historians and philosophers ask, and the very specific places where a fertile dialogue between them can take place and where it cannot.[12]

Comparably, it could at first sight appear to us that adding accounts of representation in philosophy of science and aesthetics could give us a more complete view of the same subject matter. And this is probably correct to a certain extent, but only as long as there is awareness of the underlying worries that distinguish the debate in each of these fields. Having a common subject matter (the problem of representation) might sometimes obscure the fact that motivations for addressing that matter can vary in substantial ways. For instance, philosophers in aesthetics usually identify the problem of representation with the problem of depiction, and they predominantly focus on the analysis of how we perceive pictures (Kulvicki, 2006: 535). Meanwhile, philosophers of science have been interested in how representations of many kinds – images, graphs, diagrams – are used to present data, reason with it and lead to new discovery: they have mostly advanced nonperceptual accounts of representation (Kulvicki, 2006: 536). Another example is the historical burden that the notion of 'similarity' respectively carries in aesthetics and philosophy of science, being heavier and longer in the former case than in the latter.[13] This generally implies that accounts of similarity in aesthetics tend to be sharper and more historically sensitive than in philosophy of science, where total rejection or complete endorsement seem to be the only possibilities towards the assessment of the role of similarity in representation.

At the end, there might be some irreconcilable differences between certain questions posed by philosophers of science and aestheticians. But in principle that does not invalidate the project of establishing fertile conversations between them. Much more important is to note that it is at the level of the fundamental underlying concerns where the most relevant commonalities between aesthetics and philosophy of science lie. Kulvicki has affirmed that the divergences between philosophy of science and aesthetics actually justify the attempts of dialogue about representation (Kulvicki, 2006: 536). I would add that the acknowledgement of those divergences is in itself necessary to make the dialogue fruitful. Remarkably, Kulvicki sees the benefits of the dialogue for aestheticians as well, who could develop better tools to study the phenomenon of depicting by adopting the more embracing and cognitive notion of representation that philosophers of science use (Kulvicki, 2006: 536).

Before concluding, I would like to briefly refer to two recent accounts of representation, in which the positive fruits of the integration of aesthetics and philosophy of science are especially visible, namely Bas C. van Fraassen's and Catherine Elgin's latest proposals. In these cases, the integration of elements in art and science is done in a methodologically systematic way, with clarity about the epistemic benefit of the integration and taking into consideration the singularities and concerns existing in each field.

Bas C. van Fraassen's (2008) *Scientific Representation: Paradoxes of Perspective* is one of the most comprehensive efforts to explain the problem of scientific representation in recent philosophy of science. It is, in addition, an effective attempt to connect scientific, artistic and other types of everyday representations, such as maps and caricatures. One central element in the book is the beautifully reconstructed history of the concept of 'perspective', that van Fraassen describes as thoroughly entangled in painting, geometry and technology since the Renaissance (van Fraassen, 2008: 60). The concept of perspective is introduced not merely as a metaphor or a synonym of 'point of view', but recalling the literal meaning of perspective as a 'measurement technique', in the way Leon Battista Alberti (1435) and Albrecht Dürer (1525) originally used it (van Fraassen, 2008: 8). The most important features of drawing in perspective in this view are: occlusion, systematic distortion, orientation, grain and indexical judgment (van Fraassen, 2008: 85). Van Fraassen characterizes each of these features, alluding to more contemporary discussions in aesthetics, such as Erwin Panovsky's (1945) and Martin Kemp's (1991) accounts of perspective[14], John Hyman's (2000) and Dominic Lopes' (1996) ideas on the role of occlusion in pictures and Robert Hopkins' (1998) notion of misrepresentation[15] (van Fraassen, 2008: 37).

After the detailed reconstruction of the concept, drawing on the history of art and analytic aesthetics, van Fraassen exposes the ultimate goal of his argument: to demonstrate that representing in science is also an 'art of drawing in perspective'. Moreover, occlusion, distortion and the other distinctive features of perspectival pictures are equally essential to scientific representations, even in the cases of mathematical models and other highly abstract representations. 'Descartes's analytic geometry, Newton's and Leibniz's differential and integral calculus [...] provide, on an abstract level, resources for representation so perfect that they tend to engender oblivion to the distortions on which they trade' (van Fraassen, 2008: 41). But, as with any picture drawn in perspective, these abstract representations do enclose specific distortions of their components, occlusion of many elements that are not included in the representation, selection of a particular grain (coarse or fine) and, even more importantly for van Fraassen, indications about indexical judgments that would allow us to locate ourselves in relation to the representation (van Fraassen, 2008: 66–76). In a nutshell, the careful analysis of the notion of perspective in the history of art is used by van Fraassen to disclose aspects of scientific representations that would have been otherwise overlooked. He succeeds in showing common practices of science and art, such as the activities of selecting, occluding and distorting that take place in the process of representing in the two domains.

The second proposal I would like to refer to is Catherine Elgin's latest work (2004, 2010, 2011, 2017) that develops on the tradition set by Nelson Goodman (1968). A crucial assumption underlying Goodman's and Elgin's work is that science and art share important means and goals, and derived from there, that philosophy of art as much as philosophy of science should be an integral part of the discipline of epistemology (Goodman and Elgin, 1988). In comparison to most of the approaches I have mentioned in this paper, in which concepts from art are brought into the debate in philosophy of science, the methodological postulate in this case is different: the resources of epistemology should be able to account for the cognitive achievements of both scientific and artistic representations. This idea is challenging in at least two ways. First, it demands a more flexible conception of the boundaries of epistemology. Elgin has repeatedly argued that the scope of epistemology needs to be broadened, because it is too strongly supported on a notion of 'knowledge' – as justified, true belief – that cannot explain the innumerable representations that afford epistemic access to the world without being literally true (Elgin, 1996: ix). Understanding rather than knowledge should be the object of epistemologists' concerns. And second, this idea equates the cognitive value of scientific and artistic representations. In the paper included in this volume, Elgin concludes that 'the difference between the arts and the sciences is more practical than epistemic' (this volume: 40). Works of art can be epistemically rewarding as they reorient us, enabling us to see things in the extra-aesthetic world differently from the ways we saw them before. And this is not that different from what an experiment or a thought experiment in science can do (this volume: 35). Fiction, exemplification, metaphor and depiction are means in the production of artworks, but also in the production of scientific models. The consequence of this, putting it in Goodman's words, is that 'the arts must be taken no less seriously than the sciences as modes of discovery, creation, and enlargement of knowledge in the broad sense of advancement of the understanding' (Goodman, 1968: 102).

Two different methodological lessons can be extracted from these works just described. Van Fraassen's proposal is an effective attempt to incorporate elements from art into the debate of scientific representation. It demonstrates that the dialogue with aesthetics and the history of art can be beneficial for the understanding of scientific representations, insofar as it is done in a systematic way, respectful of the tradition of debates that is characteristic of each discipline. Goodman and Elgin go one step further. Their proposals show more profound epistemic and cognitive links between science and art than probably any other contemporary attempt to connect the two domains. A specific methodological approach is present here: the problem of representation must be discussed within the domain of epistemology. Moreover, epistemology, as a normative discipline, should be able to account for the varied – and often nonverbal – means through which scientific and artistic representations succeed in affording understanding about the world. These proposals exemplify two ways of bringing philosophy of science and aesthetics together, and show that the problem of representation can be more fruitfully addressed in the intersection of the two domains. But they also show that a well-reasoned methodological strategy and explicit epistemic goals are required to justify an interdisciplinary approach of this kind.

4 Concluding remarks

In their introduction to the volume *Beyond Mimesis and Convention: Representation in Art and Science* (2010), Roman Frigg and Matthew Hunter defended the positive effects of a more thorough dialogue between philosophers of science and aestheticians. They described the way in which philosophers in each field feel the need to incorporate resources from the other field as inevitable 'covert acts of kleptomania' (Frigg and Hunter, 2010: xvi). I believe this expression captures the unavoidable act of recognizing the strengths and potential of other domains to address particularly complex problems like representation. Still, I hope to have shown why thinking in terms like 'kleptomania' is probably not the most adequate way of addressing the integration of philosophy of science and aesthetics. Stealing resources and concepts from other fields typically involves the misuse of the original ones. If our acts of incorporating elements from other traditions of thought are, instead of 'covert', explicit, noticeable, openly problematized, the results would possibly be more fruitful and precise in methodological terms.

I tried to summarize, in Section 2 of this chapter, various attempts in contemporary philosophy of science to incorporate elements from pictorial arts into accounts of scientific representation (Suárez, 2003; Chakravartty, 2010; French, 2003). Although these works have advanced significant considerations to the problem of representation and highlighted relevant commonalities between scientific and artistic products, some limitations concerning the precise role of artistic elements in their accounts were pointed out. In Section 3, I offered some general methodological comments on the project of integrating aesthetics and philosophy of science, specifically taking reflections in iHPS on interdisciplinarity as an inspiring example for it. The argument here needs to be further developed, but two recent proposals from van Fraassen (2008) and Elgin (2010, 2011, 2017) were presented as examples of particularly insightful ways of connecting scientific and artistic representations.

Notes

1 Literature on fictionalism draws on the idea that modelling practices in science are practices of fiction-making and/or concern fictive entities. The edited volumes *Fictions in Science* by Suárez (2009) and *Beyond Mimesis and Convention* by Frigg and Hunter (2010) contain a good sample of works in this direction. In this paper, I will mainly discuss how philosophers of science incorporate examples and concepts from *pictorial arts*, and will not specifically refer to the use of literary fiction to address the problem of scientific representation.

2 Suárez has recognized in several occasions that similarity might be one of the possible *means* of representation (Suárez, 2003: 230; 2004: 768; 2010: 95). His inferential view can be taken as a development of the idea of the *means* of representation, as inferential capacity can help define the sense in which representations can be accurate.

3 See for instance: van Fraassen, 2008; Suárez, 2003; French, 2003; Hughes, 1997; Contessa, 2007; Frigg, 2006; Toon, 2012.

4 This relates to Suárez's distinction between the 'means' and the 'constituents' of representation (Suárez, 2003: 230). Greenberg (2013) points out the different ways in which

these two different questions about representation have been phrased in contemporary aesthetics. The question about the 'means' of representation can be identified with the question about the singularities of 'pictorial representation', about the '*accuracy* conditions for pictorial representation' and about 'pictorial content'. On the other side, the question about the 'constituents' of representation can be identified with the questions about the 'conditions for representation' and 'pictorial reference' (reference is different from denotation in Greenberg's sense) (Greenberg 2013: 222n).

5 Dominic Lopes (1996; 2005) refers to these two conceptions as 'objective similarity accounts' and 'subjective or internalized similarity accounts'. In the objective accounts there are similarities between pictures' design (visible properties of the surface of the picture) and the properties of the subjects they represent. And in the subjective accounts our experiences of pictures are experiences of resemblances between designs and the visual field representations of the depicted scenes (Lopes, 1996: 20; and Lopes, 2005: 43).

6 'Visual field' is understood here as the abstraction of the way the world is represented in ordinary perceptual experience lacking the third dimension, like if a plane was interposed between scene and viewer (Budd, 1993: 221).

7 French tries to highlight divergences between scientific and artistic representations at this point. In the example of the Lorentz transformations, French only needs isomorphism to claim that the marks in the sand represent the natural phenomenon, while in a supposed case of marks looking like a face, he would not say that they represent a face unless there is also a clear intention causing it (French, 2003: 1473). Then, it is not completely clear why French decides to refer to this account in aesthetics in his argument once he recognizes significant difference between representation in art and science. In any case, this conception of scientific representation, that does not require the presence of agents or intentions to exist, has been challenged by numerous contemporary philosophers of science (Suárez, 2015; van Fraassen, 2008; Contessa, 2007; Callender and Cohen, 2006; Giere, 2004).

8 On the specialization of academic disciplines, and particularly of the branches of philosophy, see: Collins, Randall (1998) *The Sociology of Philosophies: A Global Theory of Intellectual Change*. Harvard University Press (specifically Chapter 12: 'Intellectuals take control of their base: The German university revolution' and Chapter 13: 'The post-revolutionary conditions: Boundaries as philosophical puzzles') and Schaffer, Simon (2013) 'How disciplines look'.

9 Illari focuses on literature on causality in philosophy of science. She describes how some philosophers use toy examples, others refer to actual examples in science and others introduce elaborate case studies to argue for their accounts. Problems arise when they don't recognize the different scope their respective proposals have. For instance, someone might place disproportionate emphasis on problems with the use of a toy example as if they were put forward as central case studies in actual scientific research (Illari, manuscript).

10 The Boston Studies' volume *Integrating History and Philosophy of Science: Problems and Prospects* (Mauskopf and Schmaltz, 2012) offers a good overview of the state of the art of the field of iHPS after fifty years of discussions.

11 Whether philosophical claims have an ahistorical character or not is another point of discussion in iHPS. Authors like Kuukkanen (2015: 1) argue that philosophy and history entail incompatible metaphysics, i.e. essentialist versus historicist metaphysics, while others like Chang (2012) claim that it is possible to conceive philosophy as a historically-engaged endeavour.

12 Dear proposes the concept of 'epistemography' to his purpose: 'I suggested that the core of [history, philosophy and social studies of science] should be seen as "epistemography" – the attempt to give an empirical account of knowledge-practices. This seems to be a useful pragmatic stance to take, even despite the obvious objection that one person's empirical account might be somebody else's distorted misrepresentation' (Dear, 2012: 71).

13 See Stephen Halliwell's *The Aesthetics of Mimesis: Ancient Texts and Modern Problems* (2002) for a historical enquiry about the origins and development of the prejudices surrounding the notions of mimesis and similarity.

14 Van Fraassen refers to Albrecht Dürer's treatise, *Unterweysung der Messung* ('Art of Measurement'), through Erwin Panovsky (Panovsky, Erwin (1945) *Albrecht Dürer*. Princeton University Press); and to Alberti's *De Pictura* (1435, 'On Painting') through Martin Kemp's introduction to Alberti, Leon Battista (1991) *On Painting*. London: Penguin Books.

15 See: Hyman, John. (2000). 'Pictorial Art and Visual Experience'. *British Journal of Aesthetics*, 40; Lopes, Dominic. (1996). *Understanding Pictures*. Oxford: Clarendon Press; and Hopkins, Robert. (1998). *Picture, Image and Experience*. Cambridge: Cambridge University Press.

References

Ambrosio, C. (2013). 'Iconic Representations, Creativity and Discovery in Art and Science', in González, W. J. (ed.) *Creativity, Innovation, and Complexity in Science*. A Coruña: Netbiblio.

Arnheim, R. (1962). *The Genesis of a Painting: Picasso's Guernica*. Berkeley, CA: University of California Press.

Budd, Malcolm. (1993, 2008). 'How Pictures Look' in *Aesthetic Essays*. Oxford: Oxford University Press.

Callender, C. and Cohen, J. (2006). 'There Is No Special Problem about Scientific Representation'. *Theoria*, 55: 5.

Chakravartty, A. (2010). 'Truth and Representation in Science: Two Inspirations from Art', in Frigg, R. and Hunter, M. (ed.) *Beyond Mimesis and Convention: Representation in Art and Science. Boston Studies in the Philosophy of Science*. London: Springer.

Chang, H. (2012). 'Beyond case-studies: History as philosophy', in Mauskopf, S. and Úmd Schmaltz, T. (eds.) *Integrating History and Philosophy of Science: Problems and Prospects. Ûoston Studies in the Philosophy of Science*. London: Springer.

Chipp, H. (1988). *Guernica*. London: University of California Press.

Collins, R. (1998). *The Sociology of Philosophies. A Global Theory of Intellectual Change*. Cambridge, MA: Harvard University Press.

Contessa, G. (2007). 'Scientific Representation, Interpretation, and Surrogative Reasoning'. *Philosophy of Science*, 74 (1): 48–68.

Dear, P. (2012). 'Philosophy of Science and Its Historical Reconstructions', in Mauskopf, S. and Schmaltz, T. (eds.) *Integrating History and Philosophy of Science: Problems and Prospects. Boston Studies in the Philosophy of Science*. London: Springer.

Downes, S. (2009). 'Models, Pictures, and Unified Accounts of Representation: Lessons from Aesthetics for Philosophy of Science'. *Perspectives on Science*, 17 (4):417–28.

Elgin, C. (1996). *Considered Judgment*. Princeton, NJ: Princeton University Press.

Elgin, C. (2004). 'True Enough'. *Philosophical Issues*, 14 (1):113–31.

Elgin, C. (2010). 'Telling Instances', in R. Frigg and M. C. Hunter (eds.) *Beyond Mimesis and Convention: Representation in Art and Science*. Berlin and New York: Springer.

Elgin, C. (2011). 'Making Manifest: The Role of Exemplification in the Sciences and the Arts'. *Principia*, 15 (3): 399–413.

Elgin, C. (2017). 'Nature's Handmaid, Art', this volume.

Elliott-Graves, A. and Weisberg, M. (2014). 'Idealization'. *Philosophy Compass*, 9 (3): 176–85.

van Fraassen, B. C. (2008). *Scientific Representation. Paradoxes of Perspective*. Oxford: Oxford University Press.French, S. (2003). 'A Model-Theoretic Account of Representation (or, I Don't Know Much About Art... But I Know It Involves Isomorphism)'. *Philosophy of Science*, 70: 1472–83.

Frigg, R. (2006). 'Scientific Representations and the Semantic View of Theories'. *Theoria*, 55: 49–65.

Frigg, R. and Hunter, M. (eds.) (2010). *Beyond Mimesis and Convention. Representation in Art and Science*. Boston Studies in the Philosophy of Science, London: Springer.

Giere, R. (2004). 'How Models Are Used to Represent Reality'. *Philosophy of Science*, 71: 742–52.

Goodman, N. (1968). *Languages of Art*. Indianapolis: Hackett.

Goodman, N. and Elgin, C. (1988). *Reconceptions in Philosophy and Other Arts and Sciences*. Cambridge, MA: Hackett Publishing.

Greenberg, G. (2011). *The Semiotic Spectrum*. PhD Thesis, Rutgers University.

Greenberg, G. (2013). 'Beyond Resemblance'. *Philosophical Review*, 122 (2): 215–87.

Halliwell, S. (2002). *The Aesthetics of Mimesis: Ancient Texts and Modern Problems*. Princeton, NJ: Princeton University Press.

Hopkins, R. (1998). *Picture, Image and Experience*. Cambridge: Cambridge University Press.

Hughes, R. I. G. (1997). 'Models and Representation'. *Philosophy of Science*, 64: 325–36.

Hyman, J. (2000). 'Pictorial Art and Visual Experience'. *British Journal of Aesthetics*, 40: 21–45.

Illari, P. (manuscript). 'Cases and Examples in the Philosophy of Causality'. Available at: https://www.academia.edu/32766275/Cases_and_Examples_in_the_Philosophy_of_Causality.

Kemp, Martin. (1991). 'Introduction', in Alberti, L. B. (1991). *On Painting*. London: Penguin Books.

Kulvicki, J. (2006). 'Pictorial Representation'. *Philosophy Compass*, 1 (6): 535–46.

Kuukkanen, J-M. (2015). 'Historicism and the Failure of HPS'. *Studies in History and Philosophy of Science*, 55: 1–9.

Lopes, D. (1996). *Understanding Pictures*. Oxford: Clarendon Press.

Lopes, D. (2007). *Sight and Sensibility. Evaluating Pictures*. Oxford: Oxford University Press.

Mauskopf, S. and Schmaltz, T. (eds.) (2012). *Integrating History and Philosophy of Science: Problems and Prospects. Boston Studies in the Philosophy of Science*. London: Springer.

Oppler, E. (1988). *Picasso's Guernica*. New York and London: Norton & Co.

Panovsky, Erwin. (1945). *Albrecht Dürer*. Princeton: Princeton University Press.

Peacocke, C. (1987). 'Depiction'. *The Philosophical Review*, 46 (3): 383–410.

Schaffer, Simon. (2013). 'How Disciplines Look', in Barry, A. and Born, G. (eds.) *Interdisciplinarity. Reconfigurations of the Social and Natural Sciences*. London: Routledge.

Suárez, M. (1999). 'Theories, Models and Representations', in Magnani, L., Nersessian, N. and Thagard, P. (eds.) *Model-Based Reasoning in Scientific Discovery*. Dordrecht: Kluwer.

Suárez, M. (2003). 'Scientific Representation: Against Similarity and Isomorphism'. *International Studies in the Philosophy of Science*, 17 (3): 225–44.

Suárez, M. (2004). 'An Inferential Conception of Scientific Representation'. *Philosophy of Science*, 71: 767–79.

Suárez, M. (2009). *Fiction in Science: Philosophical Essays on Modelling and Idealization*. New York: Routledge.

Suárez, M. (2010). 'Scientific representation'. *Philosophy Compass*, 5 (1): 91–101.

Suárez, M. (2015). 'Deflationary Representation, Inference and Practice'. *Studies in History and Philosophy of Science*, 49: 36–47.

Toon, A. (2012). *Models as Make-Believe: Imagination, Fiction and Scientific Representation*. London: Palgrave Macmillan.

2 Nature's handmaid, art[1]

Catherine Z. Elgin

1 Introduction

Works of fiction are either false or truth-valueless. Representational paintings and sculptures, being non-propositional, make no claims. Absolute music does not even denote. Whatever the value of such works, it may seem, that value is not, even in part, epistemic. In this respect, art is the polar opposite of science. I disagree. My thesis is that art, like science, embodies, conveys, and often constitutes understanding. I am not using 'understanding' in a loose or extended sense. Rather, my contention is that the arts and the sciences use the same symbolic resources to achieve much the same symbolic ends. The typical response is, 'That's ridiculous! Everybody knows that art aspires to beauty and science aspires to truth. They're not even trying to do the same thing, much less succeeding.' This stereotype is false and misleading. It distorts both the ends and the means of both disciplines. To establish this, I'll set aside worries about beauty and focus on truth.

The stereotype assumes that to function cognitively is to be or aspire to be true. But science unblushingly uses models, idealizations, and thought experiments that are not, and do not purport to be true. Physics talks of frictionless planes. But friction is ubiquitous. There are and could be no frictionless planes. To exclude friction from the description of a moving body is to represent things as they are not. Thermodynamics appeals to the ideal gas. The ideal gas is supposed to consist of dimensionless, perfectly elastic, spherical molecules that exhibit no mutual attraction. But every material object has some dimensions; no molecule is a sphere; and every object, being subject to gravity, attracts every other object. To be sure, there are more 'realistic' gas models that do not depart so far from reality. But thermodynamics must simplify. A representation of the interactions of gas molecules as they really are would be computationally and conceptually intractable. We could make no sense of it. What simplifications to make is subject to negotiation; whether to simplify is not. Scientists rely on thought experiments. Einstein asks what a person riding on a beam of light would see. The conditions he imagines could not be met. No one could ride on a light beam; it is too small. Even if, à la *Alice in Wonderland,* someone circumvented that difficulty, light travels so fast that the rider would be promptly smeared into oblivion. And even if she managed somehow to escape that fate, since her eye would be smaller than a photon, she would not be able to see anything! Still, scientists deploy such devices. They know full well that the conditions cannot be realized. But they find it epistemically fruitful to bracket the difficulties and represent things in terms of them.

It won't do to insist that such representations are merely heuristic. There is no expectation that science could or should do without them. An ideal gas law is required by thermodynamics; the Hardy-Weinberg formula is critical to population genetics; and every version of quantum mechanics has to worry about Schrödinger's cat. Moreover, such devices are not eliminable, even at the end of inquiry. To be sure, scientists expect current models and idealizations to be supplanted in the way that the van der Waals equation and the virial equation supplanted (or perhaps just augmented) earlier ideal gas laws. But current models will be supplanted by better models and idealizations – that is, models and idealizations that better perform the desired epistemic functions – not by the unvarnished truth. Or, anyway, that's what working scientists think. Since they are the ones whose decisions determine the direction in which science progresses, we should take their views seriously.

Still, it is not the case that every false representation is scientifically estimable. We have to see what qualifies a representation for scientific standing. The point here is that neither the contention that science's contribution to understanding consists of established truths, nor the contention that, even if current science falls short, its ultimate objective is a network consisting exclusively of established truths is faithful to scientific practice. Some (such as Nagel (see 1986)) might insist that philosophers know better than scientists what science should be doing. But that seems arrogant, particularly since science, as scientists do it, is an extraordinarily admirable human achievement. We seem then forced to recognize that science is epistemically estimable, despite its divergence from strict, literal truth. That being so, we have no reason to maintain that its divergence from truth excludes art from the epistemic realm.

Turn now to art. It is not unusual to emerge from an encounter with a work of art – a painting, a performance, a novel – thinking that we learned something. And what we think we learned is not just something about that work or its author – not just that Shakespeare had a big vocabulary, or that Whistler could achieve amazing effects by modulating shades of grey, or that Bach was really adept at the fugue – but something, maybe something significant, about the extra-aesthetic world. I think we are right to say such things. But that raises an important epistemological problem, for we gained few, if any, justified true beliefs about the extra-aesthetic world. Nor would we attempt to express what we have learned by articulating newly acquired, justified true beliefs. What exactly are we saying? What entitles us to say it?

Very roughly, I contend, epistemically rewarding works of art reorient us, enabling us to see things differently from the ways we saw them before. This needs to be spelled out. To do so, I will highlight some parallels between art and science. Experiments and thought experiments, I will urge, parallel fiction.

Static models parallel drawings. Dynamic models and simulations parallel performances. All function via exemplification.

2 Exemplification

Exemplification is the referential relation by means of which a sample, example, or other exemplar refers to some of its properties (Goodman, 1968, Elgin, 1996). An exemplar highlights, displays, or makes manifest some of its properties by both instantiating and referring to those properties. Indeed, it refers via its instantiation of those properties. A swatch of herringbone tweed can be used as a sample of herringbone tweed. It is an instance of the pattern that refers to that pattern. A swatch of seersucker, not being herringbone tweed, cannot serve as a sample of herringbone tweed. A sample does not exemplify all of its properties. It can highlight some of its properties only by marginalizing or downplaying others. In its standard use, a fabric sample does not exemplify its shape, age, or origin. Exemplification is selective. In different contexts, the same object can exemplify different properties. Although they are not exemplified in a tailor's shop, the size and shape of the tweed sample might be exemplified in a marketing seminar, where the focus is on what features make a commercial sample effective.

My emphasis on commercial samples may make exemplification seem narrow. It is not. An object can in principle exemplify any property it has. So exemplification is versatile. Moreover, we can turn anything into an example simply by treating it as such. All we need to do is point something out and use it to highlight a property. Exemplification thus lends itself to intellectual opportunism. A naturalist can point out a plant as an example of, say, poison ivy. He might simply point to the plant. Or he might go on to identify its important properties, perhaps the shape and sheen of the leaves, the trifoliate leaf configuration, the capacity to cause a rash. If he is successful, his audience will now know how to identify poison ivy. He has given them epistemic access to it.

In this case, exemplification was effected by pointing out properties that were easily seen. Sometimes, however, the properties we seek to exemplify are subtle or are ordinarily overshadowed by more prominent features. Then considerable stage-setting may be needed to bring the relevant properties to the fore. This is what happens in experiments. Let's look at an example drawn from middle-school science. To ascertain whether water conducts electricity, a science student does not attempt to induce a current in a vat of pond water or rain water or tap water. Such liquids contain impurities. If she were to detect a current in such a liquid, she would not be able to tell whether it was the water or the impurities that conducted electricity. Rather, she uses distilled water – water from which, as far as she can tell, all impurities have been eliminated. Then, in case the conductivity is slight (as it turns out to be), she runs her result through an amplifier. The result exemplifies water's capacity to conduct electricity. For our purposes, the critical point is that she distances herself from nature, operating on a substance (distilled water) not to be found in nature, subjecting it to a force (the amplification) greater than what is found in nature, in order to discover something about nature. Her result exemplifies something about not only pure, distilled water but also about natural fluids, such as pond water and rain water, that are, in part, H_2O.

Because an item can in principle exemplify any feature it instantiates, the range of features it has the capacity to exemplify is vast and heterogeneous. Haydn's *Farewell Symphony* brings this out. It is famous for its finale. Each year, Count Esterhazy, Haydn's employer, removed his court, including his orchestra, to his remote summer palace. In 1772 he extended their stay, seemingly endlessly. The musicians desperately wanted to return home to their wives. Haydn's *Symphony 45* made their case musically. In the finale, as each musician finished playing, he got up and left the stage. The Count got the point. The following day, everyone was permitted to return home.

Manifestly, the finale is an exemplar of the power of collective action. It could feasibly serve as the theme song for the American Federation of Musicians. It is also something of a joke, highlighting and making salient our expectation that, even if they have nothing further to play, orchestra members will remain on stage until a piece ends. But the finale is not just an editorial comment or a joke tacked onto the end of the symphony; it is an integral element of the work. *The Farewell* is, as far as we know, the only eighteenth-century symphony in F# minor. F# minor is recognized as a difficult key, and the piece can be interpreted as exemplifying the difficulties. It is not a key in which one could feel musically at home. Further, the symphony exemplifies incoherence, instability, imbalance extended over time. Tonal relations are at odds with each other and with cadences. Chord progressions go unresolved. Sonic and rhythmic patterns fail to mesh. The piece thus exemplifies tempestuousness, chaos, a forward trajectory through a musical maelstrom toward an unseen but hoped for haven. There is a strange, beautiful, D major interlude, where the basses drop out and the music seems to float, unmoored. It exemplifies a promise of coherence, but that coherence is never achieved. The listener is left hanging. The journey continues. In the end, the various musical elements are brought into accord. But the resolution is itself strange. There is no grand finale – no victory fanfare. As one by one the musicians depart, the music grows softer, until only the two violins, playing ever more *pianissimo*, remain. The rest is silence. On one reading, the symphony thus suggests that we live our lives in F# minor. With instabilities to balance, perils to traverse, incoherences to resolve, we seek a safe haven with no assurance of finding one this side of the grave. On another, it is a musical parallel to *The Odyssey*. Like Odysseus, we travel through tempestuous seas seeking not just a lovely interlude, but home (Webster, 1991). A single symbol can exemplify any and many of its features, enabling the interpreter to forge a variety of epistemically valuable connections across a variety of domains. Background information (about the key of F# minor, about *The Odyssey*) may inform an interpretation. If so, that background information is crucial to the individuation and identification of the exemplars that interpretation draws on. Still, with the appropriate background, the insights are there to be found. And with different backgrounds, different features will be exemplified and different insights will emerge.

Because they refer to some of their properties, if properly interpreted, exemplars afford epistemic access to those properties. By attending to an appropriate exemplar, we can learn to recognize herringbone tweed, poison ivy, water's conductivity, the ubiquity of unresolved tensions in human lives.

It might seem from my discussion so far that exemplification is simply a device for making manifest what is already known by some other means. If so, it would be a vehicle for conveying insights, not for generating them. But, exemplification is not always a matter of making manifest antecedently known properties. As we have seen, an experiment is designed to make manifest features that are not antecedently known. Some sample-taking functions similarly. A mining inspector collects a variety of samples in order to make manifest something no one yet knows – the distribution of gases at different levels of the mine. A political poll samples public opinion about a matter no one yet knows – perhaps the weight voters attach to economic factors in the upcoming election.

Something similar occurs in the arts. Merce Cunningham and John Cage were frequent collaborators whose method of collaboration strikes many as odd: Cunningham choreographed a dance while Cage independently wrote the music. All they agreed about in advance was the duration of the piece. Until the first performance, no one (not even the dancers, the choreographer, or the composer) knew what the work would exemplify. In some of their works and some passages of those works, music and choreography were mutually reinforcing; in others they pulled in different directions. Cage and Cunningham sought to exemplify the autonomy of the different arts. Even when writing for dance, Cage was not willing to make music subservient to the ends of dance. Nor was Cunningham willing to make dance a mere visualization of music. In cases like these, just what would be exemplified was not antecedently known. In other cases, works exemplify features that are at odds with what their authors intended. *Paradise Lost* was supposed to justify the ways of God to man. It does not succeed. Rather, it glorifies Satan, thereby exemplifying the importance of personal autonomy, self-respect, and not giving into bullies. It fails to exemplify what Milton sought to bring out.

3 Fictions in science and art

An experiment is no mere matter of bringing nature indoors. It is a controlled manipulation of events, designed and executed to make some particular phenomenon salient. Natural entities are multifaceted. Important properties and relations are often masked by the welter of complexities that embed them. In experimenting, a scientist isolates a phenomenon from many of the forces that typically impinge on it. To the extent possible, she eliminates confounding factors. She holds most ineliminable factors fixed, effectively consigning them to the cognitive background of things to be taken for granted. This enables the effect of the experimental intervention on the remaining variable to stand out. This strategy enables her to cast into bold relief factors that might typically be hidden from view.

Suppose a population of wild mice who were accidentally exposed to bisphenol-A subsequently exhibited a high rate of liver cancer. To conclude that exposure to bisphenol-A caused their disease would be premature. Those mice might have been peculiarly susceptible to liver cancer, or been exposed to a carcinogen that scientists overlooked. To glean direct, non-anecdotal evidence of a connection between exposure to bisphenol-A and liver cancer, scientists place genetically identical mice in otherwise identical environments, exposing half of them to massive doses of the chemical while leaving the rest unexposed. The common genetic endowment and otherwise identical environments neutralize a multitude of genetic and environmental factors currently believed to influence the incidence of cancer. This blocks rival explanations that might be proposed for the elevated rate of cancer in the wild population. If the exposed mice show a significantly higher incidence of cancer than the controls, the experiment exemplifies a difference that correlates with exposure to bisphenol-A.

The result of the experiment exemplifies the difference (if any) in the incidence of liver cancer between the two groups of mice. It not only instantiates the difference, it also highlights that difference. If the difference is statistically significant, then the result exemplifies a correlation between exposure to bisphenol-A and the incidence of liver cancer. Although correlation does not imply causation, a robust correlation is often evidence of causation. In this case, the background assumption that moves us from a mere correlation to a causal judgment is the well-founded conviction that the experiment was so rigorously designed and executed that nothing but the exposure to bisphenol-A could have caused the difference. That being so, the result may also exemplify a causal relation.

So far, we are just talking about the particular mice in the experiment. But the goal of the investigation is not primarily to discover their medical fates. It is to use their medical fates to learn something more general. Since the mice in the experiment were chosen arbitrarily from the class of mice with a particular genome, it is straightforward to project to other mice of the same strain. The experiment then also exemplifies the increased propensity of mice of that strain to develop liver cancer when exposed to bisphenol-A. Moreover, the mice function as model organisms, so there is reason to think that what holds for them also holds for the organisms they serve as models for – in this case, mammals, including humans. The inference from mice to mammals is a big jump. But if the background assumptions legitimating treating the mice as model organisms are sufficiently accurate and adequate, it is reasonable to treat the experiment as exemplifying a causal connection between exposure to bisphenol-A and cancer in mammals in general.

Thought experiment involves further distancing. Sometimes the imaginative rehearsal reveals that an actual experiment *need not* be carried out. The mental run-through itself discloses the relevant information. Even without physical implementation, Galileo's thought experiment discredits the Aristotelian contention that the rate at which bodies fall is proportional to their weight. Imagine a composite object consisting of a boulder tethered to a pebble. Being composed of two rocks and some rope, the composite object is heavier than either rock alone. If Aristotle is right, it should fall more quickly than the boulder. But since, according to Aristotle, the pebble falls more slowly than the boulder, once the two are tied together, the pebble should retard the boulder's fall. It should serve as a brake. Hence the rate at which the composite object falls should be between that of the boulder and that of the pebble. The composite object cannot fall both more quickly and more slowly than the boulder, so the Aristotelian commitments are inconsistent. By exemplifying the inconsistency, Galileo's thought experiment demonstrates that the Aristotelian account cannot be correct.

Sometimes an actual experiment of the sort envisioned *cannot* be carried out. It is impossible or impracticable. By imagining a person's experience while riding in a uniformly accelerating elevator in the absence of a gravitational field and his experience at rest in the presence of a gravitational field, Einstein shows the equivalence of gravitational and inertial mass. To actually run the experiment would require placing an unconscious subject in a windowless enclosure, sending him to a region of outer space distant from any significant source of gravity, restoring him to consciousness, and querying him about his experiences. This is morally, practically, and physically unfeasible. Still, the recognition that we cannot do a real experiment does not by itself legitimate stopping short. Sometimes, the infeasibility of performing an experiment translates into the infeasibility of finding out a particular fact. The reason Einstein's thought experiment is effective is that it takes the form of a challenge: suppose the specified conditions were met. How could a subject tell whether he was in one situation or the other? What evidence could he draw on? If our best efforts to identify a way to tell the difference fail, and fail for scientifically principled reasons, we have evidence of the equivalence. Collectively, our failures exemplify that if our theories are close to correct there is no difference to detect.

Like a scientific experiment, a work of fiction selects, isolates, and insulates, manipulating circumstances so that particular properties, patterns, and irregularities are exemplified. It may localize factors that underlie or are interwoven into everyday life, but that are apt to pass unnoticed because they are typically overshadowed by or interwoven with other, more prominent matters. This is why Jane Austen believed that 'three or four families in a country village is the very thing to work on' (Austen, 1814). The relations among the members of the three or four families are sufficiently complicated and the demands of village life sufficiently mundane that the story can exemplify something worth noting about ordinary life and the development of moral personality. By restricting herself to a few fictional families, Austen in effect devises a tightly controlled thought experiment. Drastically limiting the factors that affect her protagonists enables her to elaborate in detail the consequences of the few that remain.

Writers like Tolstoy paint on a larger canvas, so grander, more panoramic themes may be exemplified as well. Some exaggerate to bring particular features to the fore. Even if we will never encounter as pure an instance of weakness of will as we find in Pierre Bezuhov, or as pure an example of malevolence as we see in Iago, encounters with those characters enable us to recognize the more mixed cases that we see in everyday life. In effect, the representation of particular episodes, characters, and features is a literary counterpart of what Nancy Cartwright (1983) calls 'prepared descriptions' in science. Both distance us from phenomena in order to enable us to see them more clearly.

In *The Nicomachean Ethics,* Aristotle suggests that we should call no man happy until he is dead (Aristotle, 1953). He uses the example of Priam, the elderly king of Troy. Typically, students are unconvinced. Part of the problem is that 'happiness' is a less than adequate translation of 'eudaimonia'. 'Flourishing' would be better. But even then, the example is less than wholly convincing. After all, students insist, Priam evidently flourished throughout most of his life. It wasn't until the very end that things went badly for him. This is plausible. Priam is far from an ideal case. *Oedipus Rex* provides a more powerful exemplar of Aristotle's point. Unbeknownst to himself, Oedipus killed his father, married his mother, and thereby brought a plague on the city he ruled. Throughout most of his life, he seemed to be flourishing, but in fact he was not. On learning what he had done, he could not but look on his past with repugnance and disgust. He never actually flourished, even if it looked throughout most of his life as though he did. The suggestion then is that *Oedipus Rex* can be read as a thought experiment that exemplifies the plausibility of Aristotle's claim.

I've suggested that experiments, thought experiments, and works of fiction distance, isolate, and purify. They set up circumstances, sometimes quite unrealistic circumstances, and see how things play out. They devise contexts, including and omitting as necessary to bring out the features they seek to highlight. They can be vehicles for discovery.

Still, it might seem that we would do better to face the facts directly and avoid the detour through fiction. I disagree. A sociological investigation of country villages would not be preferable to looking at Austen's three or four families. Nor would a biography of a real leader with a tragic flaw be preferable to *Oedipus Rex*. The features that stand out so plainly in the fictions might well be obscured in 'real life' situations where other factors obtrude. Moreover, in factual studies, objections can always be raised. Maybe other, unrecognized factors figured in the events we seek to explain. Was it really pride and prejudice that kept the two apart, or were there unresolved issues stemming from their childhood, or genetically based propensities, or complicated family dynamics that we failed to accommodate? Does the tragic flaw the biographer targets really explain the subject's downfall, or were other causal factors at work? Maybe he never was as admirable as the biographer makes him out to be. Austen and Sophocles block such objections by simply omitting the potential complications. And by the effectiveness of their stories, they show that their omission makes no difference.

The same point holds in science. It would not be preferable to drop things from towers rather than to do Galileo's thought experiment. For one thing, there is no need. Galileo's thought experiment is, in effect, a *reductio* of the Aristotelian position, in much the way that *Oedipus Rex* is a *reductio* of the idea that someone's seeming to flourish demonstrates that he is actually flourishing. Moreover, dropping things from towers would muddy the waters by introducing factors that need to be controlled for and inviting questions about, for example, how high the tower has to be, how different the weights of the objects have to be, how accurate the timers have to be, whether we have adequately accommodated the effects of air resistance, wind, and so forth. Galileo evades these questions by simply omitting the irrelevant factors. They make no difference. And the effectiveness of the thought experiment makes their irrelevance manifest.

The same point holds for actual experiments. Simply charting the incidence of cancer in wild mice exposed to bisphenol-A would not make as compelling a case as an experiment. The experiment controls for genetic factors and accidental (that is, irrelevant) features of the environment. Because it does, the effectiveness of the experiment makes it manifest that those factors do not figure in the result.

Fictions, thought experiments, and laboratory experiments exclude factors deemed irrelevant so that they can investigate the consequences of those that remain. They can go to the limit – perhaps a limit that is nowhere to be found in nature, but that illuminates mundane cases. Ahab is a representation of the limit of obsession. He exemplifies how an obsession can dominate and destroy not just an individual life, but the lives of others who are involved with the obsessed individual. A simulation that represents molecular interactions at increasingly cold temperatures – colder than can actually be realized – discloses that very near absolute zero, electrical resistance disappears.

In some cases, distancing is more than merely useful. It is essential for the exemplification of the features of interest. The Judson Dance Theater is a postmodern dance collective whose works exemplify everyday movement. The dancers (looking like ordinary people, wearing ordinary clothes) walk, run, climb, carry a mattress. The movements are utterly mundane. This raises a number of questions: why call their performances dance? Why are we watching them? Why did we buy a ticket, given that we can see the very same behavior outside? Sally Banes maintains that such dances are devices for defamiliarization (2003). They make the familiar strange, contriving a context that prompts audiences to attend to features that they normally look right past. Carrying a mattress is actually an intricate, complex sequence of behaviors in which a person continually adjusts her position to accommodate changes in the center of gravity of the unwieldy load. It's quite remarkable, once you notice it. Normally, of course, you don't. You want to get the mattress moved. You have no interest in attending to the complex adjustments you make in your body to get the job done. Works like those the Judson Dance group performs, by moving ordinary activities into a new venue, change the context, refocus the lens, and attune us to aspects of things that we ordinarily overlook. They bring mundane activities to exemplify features that are typically ignored. Most of us would not register or appreciate the complex physical intelligence that goes into moving a mattress if they simply saw a couple of guys schlepping a mattress down the street.

The Miller-Urey experiment begins with chemicals believed to be present on Earth in pre-biotic times. It involves a sequence of chemical reactions whose ultimate output consists of organic chemicals and amino acids. It thus shows how life could have emerged from non-living matter. For the experiment to work, the chemicals – methane, ammonia, hydrogen, and water – had to be pure. Any hint of contamination would discredit the result. Moreover, to insure that no organic material was accidentally introduced in the course of the experiment, the chemical processes had to be completely isolated from the environment. The very same sequence of chemical reactions might have occurred innumerable times since the advent of life on the planet. But because those natural occurrences were not isolated from living organisms and processes, they would not show how life could have evolved. They would instantiate the sequence of chemical changes, but not exemplify it or its importance. Although – indeed, because – the experimental components and conditions were unnatural, the experiment revealed something important about the natural world.

The effect of exemplification can be Socratic. Then a work of art or science unseats our complacency by exemplifying that we do not know what we thought we knew. Nabokov's Humbert Humbert is a notoriously unreliable narrator. He exhibits no understanding of himself, the nature of his desire, or the appallingness of his conduct. He is, moreover, a distressingly convincing character. It is not hard to believe that someone could be that clueless about himself. We may be brought to wonder, if Humbert Humbert can be that deluded about his own character and behavior, how confident should we be that we understand our behavior, or that other people understand theirs? Ought we take our, or their, sincere self-ascriptions to be even presumptively correct? Once the question is raised, we may become acutely aware that we don't know how to answer it. One effect of *Lolita* is Socratic. We come to realize that we do not know what we thought we knew.

Experiments with surprising results have a similarly Socratic effect. The Michelson Morley experiment was conducted to detect ether drift. None was detected. The first, quite reasonable, response was to rethink the experiment. Was it well-designed? Was the interferometer sensitive enough to detect small effects? Eventually, however, scientists had to conclude that the result was solid. The reason the interferometer did not detect ether drift is that there is no ether drift to detect. They did not know what they thought they knew – that waves require a medium of transmission. As is the case in the Socratic dialogues, the failure of their complacent assumption had far-reaching consequences. It exemplified a serious flaw in their understanding of light.

4 Models

Scientific models are complex symbols that represent one thing as another (Hughes, 1997). A gas molecule is represented as a perfectly elastic sphere. A harmonic oscillator is represented as a spring. Similarly, caricature involves representing one thing as another. Margaret Thatcher is represented as a dragon and Churchill is represented as a bulldog. Representation-as is a complex form of reference that involves both denotation and exemplification.

Denotation is straightforward. A name denotes its bearer. A predicate denotes the objects in its extension. A portrait, being the pictorial counterpart of a name, denotes its subject. A map denotes the terrain it applies to, and the semantically significant aspects of the map denote features of the terrain. A generic picture, such as the picture of a warbler in a field guide, denotes each of the objects in its extension – that is, each of the warblers – and so forth. Denotation is the '-of' relation. Shanghai is the name *of* a city. This is a picture *of* my cat. Here is a map *of* the campus. Denotation, moreover, can be arbitrary. We can simply stipulate that a name denotes a bearer, or that a red dot denotes a city with a population of 100,000, and it does.

Some symbols that lack denotations function as denoting symbols. They perform the same syntactic functions as symbols that actually denote. They are empty names or predicates or pictures. Sometimes the lack of a denotation is an error. Scientists thought that there was such a thing as caloric – that is, thought that the term 'caloric' denoted a substance. Explorers thought there was such a thing as a northwest passage from the Atlantic to the Pacific – that is, that the phrase 'the Northwest Passage' denoted a sequence of navigable rivers. They were wrong. In other cases, however, denoting symbols are introduced with no expectation that they denote anything real. This is so for fictional symbols, like the name 'Frodo' or the map purporting to show the route to Mordor. It is also the case, I suggest, for models like the harmonic oscillator, or the ideal gas. The critical question is, what gives such symbols their content?

Such symbols are denoting symbols – indeed, often quite specific denoting symbols – even though they lack denotata. To explain how they function, Goodman introduces the distinction between a *p*-representation and a representation *of p* (1968). A representation *of* a horse denotes a horse. A horse-representation need not denote any actual horse. It can be purely fictional. In calling a representation a horse-picture, we classify it as belonging to a particular genre – a genre composed of all and only representations with a shared ostensible subject matter. The horse-picture genre consists of all and only pictures of horses. Some are pictures of actual horses – Secretariat, American Pharaoh, and the like. Others are fictional.

The introduction of *p*-representations might seem like a dodge. If so, the only way to determine what belongs to such a genre is on the basis of an antecedent classification of their referents. In that case we have to know what the picture is of in order to know what genre it belongs to. But this is simply false. We readily classify pictures as landscapes without any acquaintance with the terrain they ostensibly represent. Likewise, we readily classify maps as Middle-Earth-maps, and pictures as hobbit-pictures or Pegasus-pictures without comparing them to their denotata. We classify such representations on the basis of their relation to one another, not on the basis of their relation to their denotation.

The two devices – denotation and *p*-representation – enable us to explicate *representation-as:*

x represents *y* as *z* just in case *x* is a *z*-representation that as such denotes *y*.

In his famous portrait, Sir Joshua Reynolds represents Mrs. Siddons as the Tragic Muse. That is, he paints a picture of Mrs. Siddons that is a Tragic-Muse-representation. A physicist represents a spring as a harmonic oscillator when he uses a harmonic oscillator diagram or equation to denote the spring. Merely being both a denoting symbol and a *p*-representation is not enough. That is the force of the 'as such' requirement. Reynolds would not have represented Mrs. Siddons as the Tragic Muse if the painting just happened to be both a representation of Mrs. Siddons and a Tragic-Muse-picture. Nor would the physicist have represented the spring as a harmonic oscillator if his representation just happened to both denote the spring and be a harmonic-oscillator representation. The two functions need to be connected. Exemplification supplies the link. In representation-as, a *p*-representation exemplifies certain features and ascribes them to the denotation. A certain sort of somber dramatically characteristic of representations in the Tragic-Muse genre is exemplified by the figure in the portrait and ascribed to its subject, Mrs. Siddons. A capacity for a distinctive mode of oscillation, characteristic of members of the harmonic-oscillator genre, is exemplified in the diagram and ascribed to the spring. That there is no such thing as the Tragic Muse and no such thing as a (pure) harmonic oscillator is irrelevant. The portrait and the model function to make features of their referents accessible, by highlighting those features through fictional representations.

Where does this leave us? Experiments, thought experiments, and fictions contrive artificial situations to exemplify features that otherwise are likely to be epistemically inaccessible or overlooked. They make those features accessible and equip us to recognize them and appreciate their significance elsewhere. Models, like some works of art, represent one thing or one sort of thing as another. The representing symbol (the model, the picture, the diagram) exemplifies features that are instantiated by, but perhaps hard to discern in, the referent (the target of the model, the subject of the picture or diagram). They too make the exemplified features and their significance epistemically accessible. This, I contend, demonstrates my thesis: the arts, like the sciences, embody, convey, and often constitute understanding.

5 Coda

Still, one might wonder, is there *no* difference between the epistemic functions of the arts and the sciences? I suggest that there are at least two significant differences. But rather than pertaining to the question as to whether the two disciplines advance understanding, they concern how the disciplines do so.

According to a familiar stereotype, art is vague while science is precise. Actually, the opposite is the case. Science intentionally sets limits on precision. A measurement is accurate only to a given number of significant figures. Any disagreement beyond that point is irrelevant. We can, to be sure, increase the accuracy of our measurements, the only limitations being the sensitivity of the measuring devices we can build and the needs of our inquiry. But there is always a limit. Art, on the other hand, allows for unlimited precision. In principle, every difference in certain respects – the thickness of line, extension of the leg, the timbre of the voice – can make a difference to what a work exemplifies or represents.

The second difference has to do with the value the disciplines place on agreement. Goodman maintains that the same configuration of ink on paper would function differently if it were interpreted as an EKG readout or as a Hokusai drawing (1968). Only two features matter in the EKG: the shape of the wave and the frequency with which it repeats. Both are measured only to a given degree of precision. As a result, two cardiologists reading the same EKG are apt to agree about how to interpret it. Or, if they disagree, they are apt to be able to clearly articulate what about the pattern is the ground of their disagreement.

In the drawing, any of a vast number of features may matter: the size, shape, color, and intensity of the line; the constitution, weave, material, colors of the paper; differences, however subtle, in shading from one part of the paper to the next, and so on. As a result, two connoisseurs may disagree about how the drawing is to be interpreted, about what features denote, what features are exemplified, and to what degree of precision. Because an aesthetic symbol is relatively replete, and because every difference in certain respects can, in principle, make a difference, this disagreement may be interminable.

Moreover, in the case of scientific symbols like the EKG, which features are significant is settled in advance. A cardiologist may detect a subtle feature in the shape of the *p* wave that turns out to be important. But she would and should attach no significance to the fact that the line gets lighter over time. That only indicates that the printer's toner cartridge needs to be replaced. It is not her problem. A connoisseur, on the other hand, might notice that, for example, the weave of the paper supports or plays against the flow of the drawing. That no one earlier had thought to take the paper's weave to be significant does not discredit her discovery.

The reason for these divergences, I suggest, is this: Science places a premium on intersubjective agreement. Because scientists build on one another's findings by taking them as unquestioned premises, they want it to be determinate and determinable what those findings are. They are willing to trade off a measure of precision to get that agreement. Although artists are plainly influenced by one another, they do not proceed by taking previous works as unquestioned premises. Because artists do not build on one another's work in the same way as scientists, they do not have the same incentive to sacrifice precision. This leaves open the permanent possibility of disagreement. Possibly connoisseurs will never agree about the significance of the paper's weave. Science then strives for symbols that bear univocal interpretations, while art welcomes symbols that bear multiple interpretations.

I suggest then that the difference between the arts and the sciences is more practical than epistemic. The two disciplines take different trajectories in their pursuit of understanding, use the insights of their predecessors and colleagues in different ways, and, as a result, place different values on intersubjective support.

Note

1 Dryden, 'Annus Mirabilis', clv.

References

Aristotle. (1953). *Nicomachean Ethics*. New York: Penguin Classics.

Austen, J. (1814). 'Letter to Her Niece, Anna Austen Lefroy, September 9'. *Letters of Jane Austen*. Bradbourne Edition. www.pemberley.com/janeinfo/brabit16.html.

Banes, S. (2003). 'Gulliver's Hamburger: Defamiliarization and the Ordinary in the 1960s Avant Garde', in S. Banes (ed.) *Reinventing Dance in the 1960s*. Madison, WI: University of Wisconsin Press.

Cartwright, N. (1983). 'Fitting Facts to Equations'. *How the Laws of Physics Lie*. Oxford: Clarendon Press, pp. 128–42.

Dryden, J. (1913). 'Annus Mirabilis'. *The Poems of John Dryden*. <bartleby.com>.

Elgin, C. (1996). *Considered Judgment*. Princeton, NJ: Princeton University Press.

Goodman, N. (1968). *Languages of Art*. Indianapolis: Hackett.

Hughes, R.I.G. (1997). 'Models and Representation'. *PSA 1996*, Volume 2, pp. S325–336.

Nagel, T. (1986). *The View from Nowhere*. New York: Oxford University Press.

Webster, J. (1991). *Haydn's 'Farewell' Symphony and the Idea of Classical Style*. Cambridge: Cambridge University Press.

3 Of barrels and pipes

Representation-as in art and science

Roman Frigg and James Nguyen

1 Introduction

A flame is moving along a fuse. It reaches a tyre, which starts rolling down a slope. It reaches the ground and moves horizontally for a short while before it starts climbing a tilted balance, its speed being just sufficient to pass the mid-point. This tips the balance to the other side and the tyre rolls down again. After having gone up and down another smaller balance it hits a board that is tied to a ladder. The ladder falls, hitting another board, which kicks the tyre in the direction of an oil barrel on top of which there is a small trolley with a burning candle. The trolley starts moving and soon gets stuck under a metal grid with sparklers, which catch fire. This lights another fuse, setting off a small firework. A spark of the firework ignites a puddle of oil and so on.

This is the opening sequence of the 1987 film *The Way Things Go* by Swiss artists Fischli and Weiss.[1] In the 29-minute-long film, we see a seemingly endless sequence of events involving physical objects such as tyres, ladders, oil barrels, shoes and soap. The events are carefully arranged and subtly calibrated. They unfold according to exceptionless laws, and yet there is an element of surprise to them. The sequence of events fascinates and even creates a sense of suspense about what's next (a reviewer for *The Independent* enthusiastically reported that watching *The Way Things Go* was like watching a Hitchcock film). Yet there is no purpose, no cause, no finality and no meaning to either the events themselves or to their progression. What happens is aimless, and eventually pointless.

The movie is not just a piece of somewhat unusual entertainment. The title of the movie, *The Way Things Go*, has an unmistakably existential ring to it and can be seen as making reference to the fate of human ambition, the purpose of social struggle and the search for meaning in life.[2] In this way, the film uses the sequence of physical events to comment on the human condition. By likening life to the sequence of events in the film, it projects some of the properties of the sequence of film-events onto human life, and represents the *conditio humana* as a sequence of carefully calibrated but ultimately aimless events.[3]

Revert three decades. In 1953, the economists in the Central Bank of Guatemala set their Phillips-Newlyn machine (PN-machine) in motion, a system of pipes and reservoirs with water flowing through it.[4] US corporation Wrigley, one of the largest buyers of Guatemalan chicle gum, had announced that it would stop imports from Guatemala in protest to a recent land reform. The economists in the Central Bank were concerned about the effect that this would have on the national economy. They adjusted the machine to account for the macroeconomic conditions in Guatemala and let the machine run. They then switched the valve marked 'exports' to the 'closed' position and watched what happened. The flow marked 'income' started falling, and the water level in a tank marked 'surplus balances' rose, which in turn caused a fall in a graph marked 'interest rates'.

But how can a machine that pumps water from reservoir to reservoir provide insight into what's happening in the Guatemalan economy? The crucial factor is that the PN-machine is not just any system of pipes and reservoirs. It was built so that it implements principles of Keynesian economics if the reservoirs are interpreted as elements of an economy, such as the federal reserve and privately invested savings, and the flow of water is interpreted as the flow of money through an economy. By using this machine to study economic conditions in Guatemala, the economists take the machine to be a model of that economy, and the model ends up representing the Guatemalan economy as a Keynesian economy.

The PN-machine, a scientific model, and the artwork *The Way Things Go* have something in common: they both *represent* their respective targets (or subjects) *as* thus or so. The PN-machine represents the Guatemalan economy *as* a Keynesian economy and *The Way Things Go* represents life *as* a sequence of carefully calibrated but ultimately aimless events. The question then is: what establishes this sort of representational relationship? More specifically: in virtue of what does a scientific model or piece of art (*X*) represent a target system or subject (*Y*) as thus or so (*Z*)?

We take as our point of departure Nelson Goodman and Catherine Z. Elgin's discussions of representation-as in the context of artistic representation (Section 2). We then generalise their notion of representation-as so that it also covers scientific representations, which results in what we call the DEKI account of representation (Section 3). Throughout Sections 2 and 3 we use visual art and material models as examples. We continue by indicating how the account can be generalised to apply to non-concrete models and artworks (Section 4). Our approach is premised on the proposition that representations in art and science share essential traits, namely the ones identified in DEKI. We defend this claim against the view that representation in the two domains is fundamentally different, and submit that differences are ones of degree rather than kind (Section 5). We end by summing up our arguments (Section 6).

Two caveats are in order. First, when discussing scientific representation we mainly focus on models and only occasionally touch upon other kinds of representation (graphs and diagrams and so on). This limitation is owed to limitations of space and we do not imply that models are the only (or even most important) medium of scientific representation. Second, we only discuss models and artworks *in as far as they are representational*. Models can perform many functions beyond representation, and it goes without saying not all art is representational. The aim here is not to offer a general analysis of art and science; we only intend to analyse how models and works of art represent *when* they represent. Finally, we delve right into the account that we deem to be the most promising account of representation, namely representation-as. For a review of alternative accounts of representation see Frigg and Nguyen (2017a).

2 Goodman and Elgin's analysis of representation-as

Goodman and Elgin's (GE's)[5] notion of representation-as is composed of two essential ingredients: the distinction between something being a representation-of a *Z* and something being a *Z*-representation, and the notion of exemplification. We discuss each of these in turn, and then explain how they combine to form the complex representational relation of representation-as. We illustrate their account with their own example of a caricature showing Winston Churchill as a bulldog.

2.1 Representation-of and Z-representation

Denotation is the two-place relation between a symbol and the object to which it applies. According to GE, for *X* to be a representation of *Y* it is necessary (and sufficient) that *X* denotes *Y*, because 'denotation is the core of representation' (Goodman, 1976: 5). For this reason, denotation is 'representation-of' (Elgin, 2010: 4).[6]

A number of qualifications need to be added about this use of 'denotation'. First, denotation is usually restricted to language, where a name is understood as denoting its bearer. This restriction is neither essential nor helpful. Signs other than words can denote. A portrait can denote its subject, a photograph can denote its motif and a scientific model can denote its target system. There is nothing in the notion of denotation that would restrict it to language (Elgin, 1983: 19–35).

Second, even though proper names are the paradigmatic example of denoting expressions, denotation is not limited to these. Definite descriptions, proper names, indexical terms, sentences, pictures, graphs, diagrams and many other symbols can also denote. In particular, at least according to GE, predicates also denote: they denote all the objects in their extension (Elgin, 1983: 19; Goodman, 1976: 19). The predicate 'red' denotes all red things, and a picture of the hydrogen atom denotes all hydrogen atoms.

Viewing denotation as the core of representation may seem innocuous, but it has important consequences. If denotation is necessary for representation-of, then not all pictures represent in this way. Pictures showing Pickwick or unicorns do not denote anything, simply because neither Pickwick nor unicorns exist. Such a picture is therefore not a representation-of anything.

This seems counterintuitive and invites the following objection: if we recognise a picture as portraying a unicorn, then surely it represents something, namely a unicorn. GE respond to this objection by pointing out that we are misled by ordinary language into believing that something is a representation only if there is something in the world that it represents:

> What tends to mislead us is that such locutions as 'picture of' and 'represents' have the appearance of mannerly two-place predicates and can sometimes be so interpreted. But 'picture of Pickwick' and 'represents a unicorn' are better considered unbreakable one-place predicates, or class terms, like 'desk' and 'table'. [...] *Saying that a picture represents a so-and-so is thus highly ambiguous between saying that the picture denotes and saying what kind of picture it is.* Some confusion can be avoided if in the latter case we speak rather of a 'Pickwick-representing-picture' of a 'unicorn-representing-picture' [...] or, for short, of a 'Pickwick-picture' or 'unicorn-picture' [...] *Obviously a picture cannot, barring equivocation, both represent Pickwick and represent nothing. But a picture maybe of a certain kind – be a Pickwick-picture* [...] *– without representing anything.*
>
> (Goodman, 1976: 21–2, emphasis added)

This leads to the introduction of the notion of a *Z*-representation: *X* is *Z*-representation if it portrays a *Z*, where we use *Z* as a placeholder for the motif of a representation (for instance *Z* = unicorn). Derivatively, one can then also speak of *Z*-pictures, *Z*-statues, *Z*-paintings and so on, to emphasise what kind of *Z*-representation one is dealing with: a *Z*-picture is a *Z*-representation that is a picture, etc.

Some *Z*-representations are representations of *Z*s: Guido Reni's *Portrait of Cardinal Roberto Ubaldini* is a man-picture and it denotes a man (namely Cardinal Ubaldini). It is one of GE's crucial insights that cases like these are, if not exceptions, then certainly not the rule. In fact there is a complete disconnect between what kind of representation something is and what, if anything, it is a representation-of (cf. Goodman, 1976: 25–31). *Z*s do not have to be denoted by *Z*-representations and, vice versa, *Z*-representations do not have to denote *Z*s. This is obvious enough in the case of language: the word 'sunflower' is not a sunflower-representation, yet it is representation-of sunflowers (because it denotes sunflowers). The observation carries over to pictures. Adriaen Coorte's *Three Medlars with a Butterfly* is a butterfly-representation while being a representation-of the transformations of the soul; Lovis Corinth's *Innocentia* is a women-representation yet it represents innocence; and Sandro Botticelli's *The Birth of Venus* is a woman-representation and it is not a representation-of anything (because the goddess Venus doesn't exist). The divorce of *Z*-representation and representation-of *Z* is in no way an anomaly, contrived by the exalted imagination of unworldly artists. A lightning-bolt-representation denotes the fastest dog at the races without being a dog-representation; public restrooms aren't usually denoted by restroom-representations; and a map of Westeros is a territory-representation without being a representation-of anything.

What does it take to be a *Z*-representation? In the case of pictorial representation, this is a much-discussed issue. So-called *perceptual accounts* hold that a picture *X* portrays a *Z* if, under normal conditions, an observer would see a *Z* in *X* (Lopes, 1996). GE take a different route and explain *Z*-representation in terms of what they call genres (Elgin, 2010: 2–3; Goodman, 1976: 23).[7] Nothing in what follows depends on how this notion is unpacked, and so we keep operating with an intuitive understanding of how pictures are categorised according to what they portray. Our preferred take on this in the context of scientific models is discussed in Section 3.

2.2 Exemplification

An item exemplifies a property *P* if it at once instantiates *P* and refers to it. To instantiate *P* without referring to it is merely to possess *P*, and to refer to *P* without instantiating *P* is to represent *P* in a way other than by exemplifying it. An item that exemplifies a property is an exemplar (Elgin, 1996: 171). Straightforward examples of exemplification are the sample cards supplied by commercial paint companies. These cards instantiate various colours and refer to the colours instantiated (Elgin, 2007: 39).

Instantiation is a necessary condition for exemplification. But the converse does not hold: not every property that is instantiated is also exemplified. Exemplification is selective (Elgin, 1983: 71). The chip card exemplifies redness, but not rectangularity, or being an inch long, even though it instantiates these properties. Only selected properties are exemplified. There is nothing in the nature of an object that determines the selection; no properties are intrinsically more important than others. Turning an instantiated property into an exemplified one requires an act of selection, which usually depends on the relevant context. The same sample card can exemplify rectangularity if used in geometry class. The specifics depend on the context and the case at hand. One aspect, however, is crucial: exemplars provide epistemic access to the properties they exemplify (ibid.: 93). So to be exemplified, a property not only has to be selected; it also has to be epistemically accessible. We say that a property that satisfies these criteria is *highlighted*. These considerations can be summarised in the following definition:

Exemplification: *X* exemplifies property *P* in a context *C* if

(i) *X* instantiates *P*, and
(ii) *P* is highlighted in *C*.

P is highlighted in *C* if and

(α) *C* selects *P* as a relevant property, and
(β) *P* is epistemically accessible in *C*.

A sample card exemplifies, say, a certain shade of red because it instantiates it and, in the context of a paint shop, is selected as relevant and is epistemically accessible (a sample card too small to see with the naked eye would not exemplify red, nor would one that is used in a context in which colour is irrelevant).

Many works of art do not literally instantiate the properties they exemplify. Pictures and statues cannot instantiate properties like speed and elegance – after all, they are made of paper or bronze. GE acknowledge this and say that these are examples of *metaphorical exemplification* (Elgin, 1983: 81). A painting can literally instantiate the property of being grey; it can metaphorically instantiate sadness (Goodman, 1976: 50–2). Metaphorically instantiated properties can be exemplified in the same way in which literally instantiated properties are: by being highlighted. In the next section, we provide a development of GE's notion of metaphorical exemplification that emphasises the importance of the literally instantiated properties in grounding non-literally instantiated, yet still exemplified, properties.

2.3 Representation-as

A key insight on the way to a definition of representation-as is that *Z*-representations can, and often do, exemplify properties associated with *Z*s. A racehorse-picture can (metaphorically) exemplify speed; a ballerina-statue can (metaphorically) exemplify grace and elegance; and air-crash film can (metaphorically) exemplify engine failure. One could then say that an *X* represents *Y* as *Z* if *X* denotes *Y* and is a *Z*-representation exemplifying certain *Z*-properties. This is on the right track, but one last step is lacking: the exemplified properties have to be imputed to *Y*. Thus we arrive at the following definition of representation-as (Elgin, 2010: 10):

> Representation-As (RA): *X* represents *Y* as *Z* iff
>
> (i) *X* denotes *Y*,
> (ii) *X* is a *Z*-representation exemplifying *Z*-properties P_1, ..., P_n and
> (iii) *X* imputes P_1, ..., P_n, or related properties, to *Y*.

Consider GE's example of a caricature representing Churchill as a bulldog, where the caricature portrays Churchill as tenacious and ferocious. RA offers the following explanation of how the caricature does this. The caricature (*X*) denotes Churchill (*Y*). The caricature shows a bulldog (*Z*), and hence is a bulldog-representation. The bulldog-representation (metaphorically) instantiates a host of bulldog-properties. Among these, tenacity and ferocity are highlighted in the context in which the caricature is shown. Hence the caricature exemplifies tenacity and ferocity. Finally, these properties are imputed to Churchill himself.

We now see how *The Way Things Go* manages to represent the *conditio humana* as a sequence of carefully calibrated but ultimately aimless events. The film (*X*) denotes the *conditio humana* (which it does mainly in virtue of its title). The film shows a burning fuse triggering a tyre to roll down a slope etc. (*Z*), and hence is a burning-fuse-tyre-rolling-down-a-slope-etc.-representation. The film metaphorically exemplifies *Z*-properties: the careful calibration of events and their ultimate aimlessness. Finally, the movie imputes these to what it denotes, the *conditio humana.*

The natural suggestion would be to generalise RA to the scientific context by letting the *X* range over scientific models, and *Y* over their target systems, and *Z* over the content or character of models. This points in the right direction, but conditions (ii) and (iii) need to be further developed in a number of ways to be able to account for what happens in the case of scientific models (and indeed some cases of artistic representation, as we shall see).

3 The DEKI account

In this section we develop our preferred account of scientific representation, which for reasons that will become clear later we call the DEKI account.[8] Our account, which builds on RA, is primarily designed to handle scientific representation, but as we discuss in more detail below, the way that we develop RA into DEKI helps shed light on artistic representation as well.

The second condition of RA stipulates that *X* be a *Z*-representation. The notion of a *Z*-representation has intuitive appeal in the case of visual representations.[9] We readily categorise Pierre-Auguste Renoir's *La Première Sortie* as young-women-in-the-theatre-representation or a sequence of *Goldfinger* as car-chase-representation. But a system of pipes and reservoirs isn't classified as a Keynesian-economy-representation in the same way. On what grounds, then, is the PN-machine classified as a Keynesian-economy-representation? And this problem is not specific to the PN-machine. Lengths of plasticine are used as models of myoglobin; oval-shaped blocks of wood serve as models of ships; mice are used as models for humans; balls connected by sticks function as models for molecules; electrical circuits are studied as models of brain function; autonomous robots are used as models for insect cognition. In virtue of what does a material object become a *Z*-representation? Neither reference to visual appearance nor appeal to genres explains how these objects come to function as *Z*-representations.

A representation, *X*, is first and foremost an object with an associated set of properties: being such and such a size, being made out of such and such materials, and so on. The material constitution of a representational vehicle matters, and so we introduce a term of art to classify them; we can call them *O*-objects. As used here, '*O*' is simply a specification of what kind of thing *X* is. Derivatively, we speak of *O*-properties to designate properties that *X* has *qua* *O*-object. The PN-machine is a water-pipe-object, and having a flow of one litre of water through a certain hose per unit of time is one of its *O*-properties.[10]

O-objects are turned into *Z*-representations by *interpreting* their *O*-properties in terms of *Z*-properties. In the PN-machine the *O*-properties include the flow of water, the capacity of tanks and so on. These are then associated with economic properties: the production-flow of a commodity and a quantity of stocks, for example. More generally, let $O = \{O_1, \ldots, O_n\}$ be a relevant set of *O*-properties pertaining to *X*, and let $Z = \{Z1, \ldots, Z_n\}$ be a set of relevant *Z* properties. An *O-Z-Interpretation I*, then, is a bijective function $I : O \rightarrow Z$. If an *O*-property is quantitative (for instance, being *x* metres long), the interpretation also contains a function associating the values of the *O*-property with the values of the corresponding *Z*-property. Hence, an object becomes a *Z*-representation when its properties are interpreted in the appropriate manner. We therefore say that a *Z-representation* is a pair $\langle X, I \rangle$, where *X* is an *O*-object, and *I* is an *O-Z-Interpretation*.

We now identify scientific models with *Z*-representations in the following manner: a *model* is a *Z*-representation where *X* is an *O*-object that is used as the vehicle of the model in a certain context (either due to convention or the stipulation of a scientist, or group thereof) and *I* is an interpretation. We then write $M = \langle X, I \rangle$ and also speak of a *Z*-model. So the reservoir-and-pipe system becomes a Keynesian-economy-representation when, in a certain context, it is used as the vehicle of the model and it is endowed with an interpretation that maps its hydraulic properties to economic properties.

It is a deliberate choice that this definition of a model contains no reference to a target system. There are models that don't have target systems, and therefore we should distinguish between the notions of being a scientific model and being a scientific representation. Some *Z*-models are also representations of a *Z*; others aren't. The PN-machine is a representation of the Guatemalan economy. But Maxwell's aether-model is not a representation-of anything (there is no aether!) despite being an aether-representation. Crucially, targetless models need not be failures. In some cases, models are constructed without being intended to be representations-of systems in the world, and an account of modelling that undercuts such an enterprise gets started on the wrong foot (we return to such models in Section 5, where we also give examples).

It is worth noting that *O* and *Z*, while often distinct, can coincide. In such cases, the interpretation *I* is the identity function. The architect's cardboard house is a house-object that is used as a house-representation, and when studying ships engineers often use small ship-shaped objects as ship-shaped-object-representations. Such representations are usually considered to be *iconic models* (Black, 1962).

Models, understood as *Z*-representations, exemplify selected *Z*-properties. The PN-machine, for instance, exemplifies rising surplus balances and falling interest rates. But, just as a painting does not literally instantiate sadness, the PN-machine does not literally instantiate falling interest rates (it's a water-pipe system!). The problem is that if $O \neq Z$, then the model-object *X* does not instantiate properties associated with *Z*, and thus cannot exemplify them. It's at this point that GE rely on the notion of metaphorical instantiation: although the painting doesn't literally instantiate sadness, it does metaphorically instantiate it, and can therefore exemplify it. GE are right in pointing out that it is not necessary that X *literally* instantiates P. But rather than relying on the somewhat vague – and to some philosophically suspicious – notion of metaphorical instantiation, we turn to the notion of an interpretation to define a precise sense of non-literal instantiation. Given that an interpretation establishes a one-to-one correspondence between O-properties and Z-properties, it is natural to say that a model $M = \langle X, I \rangle$ *I-instantiates* a Z-property P iff X instantiates an O-property P', which satisfies the following condition: P' is mapped to P by I (and if the property is quantitative, the relevant value of P' is mapped to the relevant value of P).

The introduction of I-instantiation specifies precisely how objects can exemplify properties they do not literally instantiate, and it does so in a way that emphasises the importance of the properties literally instantiated by models (their O-properties) in establishing the exemplification of the relevant Z-properties. Exemplification of Z-properties only happens under an interpretation, and for this to happen a model must instantiate the relevant O-properties that the interpretation function takes to the exemplified Z-properties. Notice that all of this can be made sense of without the need to appeal to metaphorical instantiation (although those happy with the notion of metaphorical instantiation can see the notion of I-instantiation as regimenting how scientific models metaphorically instantiate properties: they do so in virtue of a combination of literally instantiating O-properties and interpretations).

I-instantiated properties can be I-exemplified if they are I-instantiated and highlighted (as described in Section 2.2). The PN-machine, then, I-instantiates falling interest rates and commodity flows while instantiating particular meter readings and flows of water, and it I-exemplifies falling interest rates and commodity flows if they are I-instantiated and highlighted.

The next question to ask is: what makes the PN-machine represent the Guatemalan Economy? Or more generally: what makes a model, construed as a Z-representation, represent a target system as a Z? For a model to represent a target as Z, two further conditions have to hold. The first is that the model denote the target system (which, as we have seen in Section 2.1, can also be a type rather than a token). Denotation is the core of representation. It establishes representation-of. Nevertheless, as we have seen above, it is only necessary and not sufficient for representation-as. This is where the second condition comes into play. The basic idea is that properties exemplified by the model are imputed to the target. Imputation can be analysed in terms of property ascription. The model user may simply ascribe the exemplified properties to the target system, and this is what establishes that the model represents the target as having those properties.

But the properties imputed are rarely exactly those exemplified by the model. The model could, for instance, exemplify being frictionless, but the property imputed to the target is something like 'having sufficiently low friction to be negligible in the current context'. In some cases the imputed properties could diverge significantly from those exemplified by the model. It is therefore crucial that the relation between them is articulated with precision. For this reason, we build an explicit specification of how the exemplified properties are related to properties imputed into our account of scientific representation by means of a 'key'. Let $P_1, \ldots, P_n$ be the *Z*-properties exemplified by the model, and let $Q_1, \ldots, Q_m$ be the properties that the model imputes to the target (*n* and *m* are positive natural numbers which can but need not be equal). Then the representation must come with a key *K* specifying how exactly $P_1, \ldots, P_n$ are converted into $Q_1, \ldots, Q_m$. Borrowing notation from algebra (somewhat tongue-in-cheek) we write the key as a function *K*, taking a set of exemplified properties as the arguments and mapping them to a set of to-be-imputed properties:

$$K(\{P_1, \ldots, P_n\}) = \{Q_1, \ldots, Q_m\}$$

P and *Q* properties often are different, but it's worth noting here that it needn't be the case that the *P* properties are mapped to distinct *Q* properties: the key can be the identity. This would allow for models to exemplify 'relevant properties' which they are hypothesised to share with their target systems, which amounts to the claims of those who defend versions of the similarity account of scientific representation (Giere, 2004, 2010; Weisberg, 2013). Moreover, since we place no restrictions on the sorts of properties that are exemplified, we do not rule out structural properties being exemplified and then imputed onto their target systems in virtue of hypothesising that there is some structure-preserving mapping that holds between the two (such as a homomorphism [Bartels, 2006], or a partial-isomorphism [Bueno and French, 2011; French, 2003]).

Gathering together the pieces we have discussed yields the DEKI account of representation:

> DEKI: Let $M = \langle X, I \rangle$ be a model, where *X* is an *O*-object that serves as the vehicle of the model and *I* is an *O*-*Z*-interpretation. Let *T* be the target system. *M* represents *T* as *Z* iff all of the following conditions are satisfied:
>
> (i) *M* denotes *T*.
> (ii) *M* *I*-exemplifies *Z*-properties $\{P_1, \ldots, P_n\}$.
> (iii) *M* comes with key *K* associating the set $\{P_1, \ldots, P_n\}$ with a set of properties $\{Q_1, \ldots, Q_m\}$: $K(\{P_1, \ldots, P_n\}) = \{Q_1, \ldots, Q_m\}$
> (iv) *M* imputes at least one of the $\{Q_1, \ldots, Q_m\}$ to *T*.

The account owes its name to the key ingredients: denotation, exemplification, keying up and imputation. Figure 3.1 demonstrates how the various aspects of the account fit together.

Understanding how these conditions are met in the case of the PN-machine illustrates how the account works. The machine (*X*) is a water-pipe-object (*O*). *Z* is a Keynesian economy. The machine is endowed with an *O*-*Z*-interpretation (*I*), mapping hydraulic properties to economic properties. The

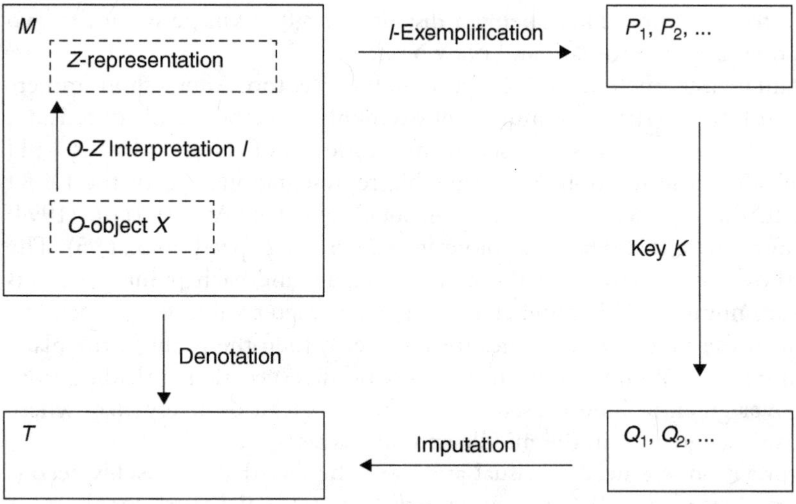

Figure 3.1 The DEKI account of representation.

machine so interpreted is a Keynesian-economy-representation, and as such it is a model M (a Keynesian-economy-model). The Guatemalan economists used M as a representation-of the Guatemalan economy by letting the model denote the Guatemalan economy (i). They did so by borrowing the denotation of the linguistic expression 'Guatemalan economy', and the model denotes whatever the term denotes. The machine instantiates a number of water-pipe-properties and, via I, it I-instantiates a number of economy properties. Some of them – the effect that a decrease in foreign exports had on income and the interest rate, for instance – are exemplified because they were highlighted (ii). We can presume that the economists used an interval-valued key, which moved from specific changes in value of the interest rate in the machine before and after the change in foreign exports to values ± 10% around them (iii) and imputed the result to the Guatemalan economy (iv).

The above-mentioned examples of models (the plasticine model of myoglobin, etc.) can be analysed along the same lines.[11] The introduction of keys was originally motivated by maps, and maps therefore (unsurprisingly) can also be analysed in terms of DEKI. A map, considered as an object, is a paper-with-colour-print-object. Under an interpretation that takes certain lines to indicate borders, blue to designate water and black dots to signify cities, the map becomes a territory-representation. Through the introduction of denotational relationships between the map and parts of the world, usually by borrowing denotation from language (by saying that the map denotes the world, that a certain dot denotes Paris, etc.), the map becomes a representation-of the world. The map then exemplifies certain properties, for instance, that the points labelled 'Paris' and 'New York' are 29 cm apart. The map comes with a key specifying scale of the map (for instance 1: 20,000,000), which translates 29 cm into 5,800 km. There being a distance of 5,800 km between the two cities is then imputed to Paris and New York.

Certain measurement devices function in this way too. After a short immersion in a solution, a strip of litmus paper exemplifies a certain shade of red and, via a key that converts a colour spectrum into levels of acidity, ascribes a pH value of 3.5 to the solution. Some graphic representations also fit the DEKI mould. In the representation of the Mandelbrod set in Argyris *et al.* (1994: 660), a key is used that translates colour into divergence speed (ibid.: 695). The square shown is a segment of the complex plane, and each point represents a complex number. This number is used as parameter value for an iterative function. If the function converges for number *c*, then the point in the plane representing *c* is coloured black. If the function diverges, then a shading from yellow over green to blue is used to indicate the speed of divergence, where yellow is slow, green is in the middle and blue is fast.

Interpretation is crucial in visual arts too. The fact that we readily recognise Edgar Degas' *The Rehearsal of the Ballet Onstage* ('*Rehearsal*' for short) as a ballet-representation may mask the fact that this recognition is the product of an interpretation. Symbolist painter Maurice Denis (1909/2003) famously reminded his fellow-artists that a painting, before being a battle horse, a nude or some anecdote, is a plane surface covered with pigments. A painting per se is a welter of lines and dots, a bounded collection of curves, shapes and colours. Assume that we make a temperature measurement at each point of a surface (for instance, the bonnet of a car) and use a colour-coding similar to the one used for the Mandelbrod set to record the outcomes in the form of a plot. Further assume that it so happens that the temperature distribution is such that the resulting temperature plot is visually indistinguishable from *Rehearsal*. Would we say that this plot is a ballet-representation? No. A coloured surface that looks like *Rehearsal* is a ballet-representation only under an interpretation that takes the colours of surface to be representations of a visual experience we have when seeing ballet dancers.[12]

Emphasising the importance of an interpretation in understanding a visual pattern is more than just an academic point. Much confusion can be avoided by bearing in mind that visual patterns are not 'natural' depictions of something just because they look like something, where 'natural' is taken to mean that there is some objective relation between the depiction and the depicted that does not depend in any essential way on the role of onlookers and observers.[13] This point is brought home by the case of Putnam's ant, which traces a line through the sand that ends up looking like Churchill (Putnam, 1981). The trace isn't a Churchill-representation, let alone a representation-of Churchill, unless it's interpreted as such. And although the visual similarity between the trace in the sand and the British politician *can* form the basis of such an interpretation (an onlooker could interpret the shape of the trace as the shape of Churchill's face with a cigar in his mouth for example), they needn't. And without an onlooker there is no interpretation to begin with, and the trace is not a *Z*-representation of any kind.[14] The adoption of an interpretation is a conventional choice and *Z*-representations don't have to be objectively related, via visual similarity or otherwise, to *Z*s (this is not to say that there never is such connection; the point is that such a relation does not turn something into a *Z*-representation without the adoption of an appropriate interpretation).

The importance of an interpretation is highlighted by considering cases where the 'obvious' or 'natural' understanding of an image is in fact not the correct one. James Elkins discusses striking cases of such images. One of his examples is a widely reproduced Hubble Space Telescope image of young stars in the Eagle Nebula (Elkins, 2007: 10–12). We see an image that looks like an underwater photograph of a rock formation that is covered with a thin layer of brownish seaweed. The unsuspecting onlooker is seduced into thinking that young stars in the Eagle Nebula look like seaweed-covered rock formations, and part of the popularity of such images derives from the seemingly easy visual access they provide to astronomical phenomena. But, as Elkins points out, this reading of the image is profoundly mistaken. The image was combined from 32 individual images taken with four different cameras. These images were cleaned, stitched together and given false colours. The colours that appear to represent an ordinary visual impression in fact are a coding for physical properties of the objects (blue, for instance, stands for the emission of doubly ionised oxygen). Unassuming onlookers unaware of all this will radically misinterpret the image.

In better cases, visual interpretations that are misleading on the surface at least raise interesting questions. Benoît Mandelbort (1982) presents an impressive collection of images that are the result of mathematical algorithms and colour codings of the kind described above, and yet look like depictions of mountains and planets, and Michael Barnsley (1993) produced a welter of images of the same kind that look like ferns. These and similar achievements were hailed as the discovery of the 'fractal geometry of nature' (as Mandelbrot calls it). It is surely remarkable that fern lookalikes can be produced by mathematical algorithms plus a colour-coding scheme, but announcement of the discovery of the fractal geometry of nature may well be premature. Per se, these images tell us more about an onlooker's interpretation than about nature itself. Filling the gap between appearance and an underlying mechanism has become the subject matter of the field of research known as fractal growth theory, which attempts to show that the equations generating the images can be seen as representations of real physical or biological processes, and therefore the shapes seen in the computer-generated images are reflective of natural process. If true, that is a significant discovery, and one that goes way beyond the superficial observation that a computer plot, when seen through a visual-image-interpretation, looks like a fern or a planet.

Returning from cautionary notes to constructive explanation, DEKI has the means to explain the working of symbolic art. Frans Pourbus the Younger's painting of Anne of Austria is, in our parlance, a princess-with-dog-representation. The painting is also a representation-of Princess Anne, because it denotes the princess. But it is not a representation-of her dog (even if she had one); the part of the painting showing a dog does not denote anything (the painting doesn't function like a portrait of a royal couple where half of the painting denotes the queen and the other half the king). But the dog is an important part of the picture and can't be dismissed as a mere ornament. The dog is exemplified. Under the conventions used at the time, the dog was a symbol for fidelity, and so the painting should be read as coming with a key associating a dog with fidelity (much in the same way in which litmus paper comes with key associating the colour red with acidity). The painting then imputes the thus keyed-up property to the princess and represents her as faithful.

4 Non-concrete objects

Not all models are physical objects, and not all artworks are visible and tangible. Issac Newton's model of the sun-earth system consists of two perfect spheres with a homogeneous mass distribution gravitationally interacting with each other but nothing else, and Leonardo Fibonacci's model of a population consists of immortal rabbits reproducing indefinitely at a constant rate, living in an environment that places no restrictions on either food or space. Mark Twain's *The Adventures of Tom Sawyer* tells a story about Huckleberry Finn and Tom Sawyer, two wayward boys exploring the Mississippi, and Louis-Ferdinand Céline's *Journey to the End of the Night* follows antihero Ferdinand Bardamu on his journeys through France and the United States.

These objects don't exist; they can't be seen and they can't be touched. They are non-concrete. They are often regarded as fictional objects or characters. How to analyse such objects is a formidable philosophical problem (indeed, there is a question already as to whether they are objects at all), and there are more options available than we can mention here.[15] For our purposes, it does not matter which options we choose. Since things like Huckleberry Finn and immortal rabbits are accessed through the imagination, we refer to them as 'imagined-objects'. The hyphen indicates that we use this locution as a term of art whose sole purpose (in this context) is to provide us with a convenient way to talk about these things while remaining ontologically non-committal. Imagined-objects can have properties. Bardamu is a gnome and Tom Sawyer is infatuated with his classmate Becky; Newton's planets are spherical and Fibonacci's rabbits are immortal. How such property attributions are analysed depends on which view of fiction one adopts.[16]

What matters for our current purposes is that imagined-objects can be interpreted in the same way in which material objects can be interpreted. Phillips and Newlyn interpreted the hydraulic properties of their machine as economic properties. Newton did the same in the case of his model of the solar system. The basic imagined-object of the model is the so-called two-body system: a system consisting of two perfect spheres with a homogenous mass distribution, one large and one small, attracted to each other with a force. In the Newtonian model, the larger sphere is interpreted as the sun, the smaller sphere as the earth and the force as gravity. So in the context of the Newtonian model, the two-body system is a solar-system-representation.

The interpretation is independent from the basic imagined-object and could in principle be changed. This is what happened in the Bohr model of the atom, which uses the same imagined-entity (the two-body system) but the large sphere is interpreted as a proton, the small sphere as an electron and the force as electrostatic attraction. Thus, in the context of the Bohr model the two-body system is a hydrogen-atom-representation.

Some works of literature can be seen as working in the same way. George Orwell's *Animal Farm* tells the story of a farm that is run by the animals themselves. But the novel is not a manifesto for the self-governance of non-humans or a demonstration of the intelligence of pigs. The novel is an allegorical denunciation of Soviet-style communism as an exploitative reign of terror. The pigs are to be interpreted as the party functionaries and other animals – horses, chicken, sheep and so on – as other segments of society; the happenings on the farm are to be interpreted as political events. Thus interpreted, *Animal Farm* is Soviet-communism-representation. As such, it need not be a representation-of any particular country or party apparatus. But in a letter to a friend Orwell described the novel as a tale against Stalin, indicating that the novel denotes Soviet Russia during the first half of the Twentieth Century, and a number of characters in the novel denote concrete historical figures: the pig called Napoleon denotes Stalin, Snowball denotes Trotsky, Squealer denotes Molotov, etc. The plot exemplifies a number of features like power being built on a cult of personality, loyalty and hard work not being rewarded, decisions being arbitrary and innocent creatures being sacrificed mercilessly in the power games of a ruthless and selfish elite. All these are imputed (with an identity key) to Stalin and his entourage, thus providing a piercing criticism of the phoney pretensions of communism.[17]

Voltaire's *Candide: or, Optimism* tells the story of a young man, Candide, who adheres to the teachings of Professor Pangloss and believes that everything in the world is for the best. But when he starts travelling the world, experiencing hardship, disaster and suffering, he becomes disillusioned with Pangloss' doctrines, which he comes to see as fundamentally at odds with how things are. On the face of it, the book is a story about the adventures of a good-hearted but naïve traveller, and the story betrays Pangloss' optimism as a doctrine that is fundamentally at odds with the course of events in the world. But we miss an important point if we stop here. Voltaire wrote the book as a response to Leibniz's doctrine that we live in the best of all possible worlds, created by a benevolent and omniscient God. In fact, Professor Pangloss is a parody of Leibniz, and so we should read Professor Pangloss as denoting Leibniz. The story exemplifies there being an unbridgeable gap between optimist teachings and real-world events, denouncing the optimist doctrine as a piece of bogus philosophy. These properties are imputed to Leibniz's philosophy (again with an identity key), and Leibniz himself is portrayed as a promulgator of a delusional and ultimately dishonest vision of the world.

These two examples aren't handpicked exceptions. Satirical and allegorical works can generally be interpreted in the same manner as the above, and so can fables and parables. Realist fiction also fits the mould (as we will see in the next section), and so do historical and biographical novels.

5 Representation in art and science

So far we have stressed the parallels between representation in art and science, and argued that both can be accommodated within the DEKI framework. This does not imply, however, that representation in art and science are identical in all respects. There are important differences. But these, we claim, are often differences of degree rather than kind. An exhaustive treatment of these differences is beyond the scope of this essay (arguably, any discussion of this issue will always remain open-ended), and so we concentrate on few focal issues: the role of targets, the flexibility of interpretation and the importance of rhetoric and style. To keep the discussion manageable, we restrict attention to literature; similar points could be made about other art forms.

A fundamental objection to the project of drawing parallels between representation in art and science is that artistic representation have no well-defined target. Writing specifically about literary fiction, Currie notes that '[w]e have no more than the vague suggestion that fictions sometimes shed light on aspects of human thought, feeling, decision, and action' (2016: 304). Since we don't find real-life analogues of, say, Natasha and Pierre (in Leo Tolstoy's *War and Peace*) we cannot compare the novel and the world, which pulls the rug from underneath the project of likening representation in art and science, because such a comparison is a central feature of scientific modelling.

The contrast between scientific models and literary fiction is rather less stark. First, not all scientific models have targets. There are famous failures like models involving the aether, phlogiston, Ptolemaic epicycles, steady state cosmology and Lamarquean inheritance of acquired characteristics. But not all targetless models are remnants of failed scientific projects. Models of three-sex reproduction in population dynamics (Weisberg, 2013), the φ^4-model in quantum field theory (Hartmann, 1995), the Lorenz model of the atmosphere (Smith, 2007), the Kac-ring model in statistical mechanics (Werndl and Frigg, 2015), the logistic model of population growth (Hofbauer and Sigmund, 1998) and the baker's model in chaos theory (Frigg *et al.*, 2016) are all models without targets. Crucially, they aren't failures. They were known all along not to have targets, and they were constructed for purposes other than the exploration of a particular target.[18] Second, not all works of literature lack targets. As we have seen above, satirical novels like *Animal Farm* and *Candide* can have clearly specified targets. Biographical novels like Vargas Llosa's *Aunt Julia and the Scriptwriter* are tales about real-world characters. Works in the tradition of social realism, such as Émile Zola's *Germinal* and Charles Dickens' *Oliver Twist*, offer piercing commentary on social reality and fierce criticism of poverty. Erich Maria Remarque's *All Quiet on the Western Front* and Kurt Vonnegut's *Slaughterhouse-Five* are passionate denunciations of the horrors of the First and Second World Wars (respectively).

One may argue that the horrors of World Wars or Stalin's cult of personality are too broad and unspecific to serve as targets. Maybe they are, and there is a discussion to be had about what counts as a target system and how it is delineated. But it pays to note that also in scientific contexts not all target systems are precisely circumscribed. Economic models represent general phenomena such as unemployment, inflation, business cycles and exposure to risk; ecologists model general processes such as population growth and predator-prey dynamics; physicists model the approach to equilibrium; sociologists model social exclusion; political scientists have models of conflict resolution. None of these are specific. Hence, if there is difference in specificity between the targets of literary fiction and scientific models, then the difference seems to be one of degree rather than kind, and the dimensions along which comparisons are made is largely uncharted territory.

The grain of truth in Currie's observation is that not all novels have even a vague target. Franz Kafka's *The Castle* or Fyodor Dostoevsky's *Crime and Punishment* are not about anything in particular, at least not in any obvious way. They are not about the Second World War or poverty. This does not mean, however, that readers cannot take the novels to be about specific things. The plaintiff trying to manoeuvre her way through the endless and often uncooperative positions of a contorted legal system may interpret *The Castle* to be about her legal nightmare; the remorseful criminal can recognise himself in Raskolnikov. The choice of a target in such cases is *ad hoc*, and a myriad of other targets are equally possible. Readers are free to choose targets, and when they do so they can use the novel to generate insights about their chosen target. It seems to be correct to say that this kind of underdetermination of targets is more common in literature than in science, but at the same time it should be acknowledged that the phenomenon is not unheard of in science either. The harmonic oscillator is the physicist's favourite workhorse, and almost anything from the atoms in the wall of a black body to insulin receptors has at some point or other been modelled as a harmonic oscillator.

A point where the difference between science and art is more pronounced is the flexibility of interpretation (in the sense of DEKI). In scientific cases, the *Z* is usually fixed by the context and the interpretation highly regimented. Someone who doesn't interpret the large sphere as the sun simply doesn't understand the Newtonian model. In literature, there is often more flexibility. How much flexibility there is depends on the context and the genre.[19] There is little flexibility in interpreting *Animal Farm* while there are (almost) no limits to an interpretation of *The Castle*. Fischli and Weiss' film we described in the introduction also lends itself to different interpretations. We interpreted it as a *conditio-humana*-representation. Someone else might emphasise the borderline functionality of the arrangement and its constant risk of failure and see it as risk-representation. Feminists might point to the masculine character of the materials and see the design of the setup as a manifestation of the male preoccupation with mechanical processes; for them *The Way Things Go* could be a gender-ideology-representation. And so on. In artistic contexts, the interpretation is often deliberately left open, and coming up with an interesting interpretation is often a creative act in its own right. Such freedom is foreign to science, where interpretations are regimented and controlled.

A last point we want to consider is the importance of rhetoric and style in the presentation of a model or a work of literature. Language and rhetoric is a crucial aspect of a work of literature. We admire great authors not only for the inventiveness of their plots, but also (and sometime even more so) for their use of language, the elegance of their expressions and the fluency of their diction. This importance of language and rhetoric, opponents of a parallelism of modelling and fiction point out, is an aspect that's entirely foreign to science. Currie submits that '[m]odels are not dependent for their value in learning on any particular formulation' (Currie, 2016: 305), while formulations are crucial in literature. A recounting of the plot of *One Hundred Years of Solitude* in the language of a seven-year-old is not the work of art that Gabriel García Márquez created.

There is no question that language and rhetoric play a different role in literature and in the presentation of scientific models, but that does not imply that models are completely independent of their formulation. Everybody who has ever spent time solving differential equations will know that the choice of the right coordinates for the description of the situation is crucial. In a recent paper discussing models (understood as imaginary entities), Vorms (2011) points out that what she calls the 'format of a representation' is crucial to the inferences scientists can draw from the model. The very same model, when presented under a different format, can yield different predictions and offer different explanations. Formulation matters. So, once again, the difference is one of degree and detail rather than kind.

6 Conclusion

The DEKI account of representation, building on Goodman and Elgin's notion of representation-as, highlights the commonalities between scientific and artistic representation. By understanding how each of DEKI's conditions are met, we come to understand how a hydraulic system like the PN-machine can represent the Guatemalan economy as a Keynesian economy, and how a cleverly calibrated sequence of rolling tyres and burning barrels can represent the *conditio humana* as ultimately aimless. The account explains, in general, how an object *X* represents a target *Y* as thus or so *Z*. This is not to say that representation-as works in exactly the same way in science as in art (or even to say that it works in exactly the same way across the sciences or across the entire field of art). DEKI's conditions are stated at the appropriate level of abstraction so that they can be met in different ways in different cases, as we have discussed. But the differences that emerge in different instances, or types of instance, of representation-as depend on how the very same conditions, of denotation, exemplification and so on, are met. We conclude by re-emphasising that our analysis is aimed at cases of scientific and artistic *representation*. We don't want to claim that all scientific models, let alone works of art, play representational roles. But where they do, we hope that analysing them through the lens of DEKI will help us understand how they work.

Acknowledgements

We would like to thank Fiora Salis for helpful discussions and comments on an earlier draft. Thanks go to Otávio Bueno and Steven French for inviting us to be part of this project.

Notes

1 A sequence of the movie can be seen at https://www.youtube.com/watch?v=GXrRC3pfLnE.
2 This is clearer in the original German title *Der Lauf Der Dinge*.
3 We briefly mention alternative interpretations in Section 5.
4 Our discussion of the Phillips-Newlyn machine draws on our (m.s). The machine can be seen in action at https://www.youtube.com/watch?v=k_-uGHWz_k0.
5 When referring to views shared by Goodman and Elgin, we use the acronym 'GE' to refer to them jointly.
6 We put systematicity above grammatical correctness when we write '*X* is a representation-of *Y*'. For a detailed discussion of GE's view on representation-of see (Frigg and Nguyen, 2017b and ms).
7 Other options are also available. For a survey, see Kulvicki (2006).
8 For more details about the DEKI account, see (Frigg and Nguyen ms).
9 This is not to say that this concept needs no further analysis; it's just to say that there is at least a pre-theoretic intuition we can build on.
10 *X* does not uniquely determine *O*. The PN-machine could also be described as a metal and plastic object, or as post-war production object. Any property instantiated by *X* could ground *O*.
11 For want of space, we cannot discuss each case individually. For useful discussions of the model of myoglobin, see (de Chadarevian, 2004), model of ships (Leggett, 2013; Sterrett, 2002), model organisms (Ankeny and Leonelli, 2011), molecules (Toon, 2011), brain functions (Sterratt *et al.*, 2011) and robots (Webb, 2001).
12 Explaining how this kind of interpretation works is no easy feat; see Kulvicki (2006) for a useful review of the options discussed in the philosophy of art.
13 Suárez (2003) emphasises this in the scientific context.
14 See French (2003), Chakravartty (2001) and Bueno and French (2011) for further discussions of this thought experiment in the context of scientific representation.
15 For reviews of these options, see Friend (2007) and Salis (2013). See also French (2010) who argues that we can adopt a 'quietist stance' towards the ontology of scientific models and theories.
16 We favour an anti-realist approach to imagined-objects and analyse property attribution as pretend instantiation; see Frigg and Nguyen (2016) for details. We emphasise that talk about imagination does not commit us to the view that thinking about models involves mental imagery; see Salis and Frigg (forthcoming).
17 An alternative analysis would take the story at face value and see the plot as an animal-farm-representation. The conversion of animal-farm-properties into Soviet-communism-properties would then be put into the key. We are not adjudicating between these options here. In our view it is a strength of the framework that it has the flexibility to accommodate different analyses of a work of literature.

18 It has been emphasised variously in the debate about models that models perform a number of functions other than representation. See Knuuttila (2005; 2011), Peschard (2011), Bokulich (2009) and Kennedy (2012) for a discussion.
19 See Eco (1992; 1994) for discussions about the limits as to how literary texts can be interpreted.

References

Ankeny, R. A. and S. Leonelli. (2011). 'What's so Special about Model Organisms?'. *Studies in History and Philosophy of Science Part A, 42 (2)*, pp. 313–23.

Argyris, J. H., G. Faust and M. Haase. (1994). *Die Erforschung Des Chaos: Eine Einführung Für Naturwissenschaftler Und Ingenieure*. Braunschweig: Vieweg+Teubner Verlag.

Barnsley, M. (1993). *Fractals Everywhere*. Boston: Academic Press.

Bartels, A. (2006). 'Defending the Structural Concept of Representation'. *Theoria, 21*, pp. 7–19.

Black, M. (1962). *Models and Metaphors: Studies in Language and Philosophy*. Ithaca, NY: Cornell University Press.

Bokulich, A. (2009). 'Explanatory Fictions', in M. Suárez (ed.) *Fictions in Science: Philosophical Essays on Modelling and Idealization*, London and New York: Routledge, pp. 91–109.

Bueno, O. and French, S. (2011). 'How Theories Represent'. *British Journal for the Philosophy of Science*, 62, pp. 857–94.

Chakravartty, A. (2001). 'The Semantic or Model-Theoretic View of Theories and Scientific Realism'. *Synthese, 127*, pp. 325–45.

Currie, G. (2016). 'Models as Fictions, Fictions as Models.' *The Monist, 99*, pp. 296–310.

de Chadarevian, S. (2004). 'Models and the Making of Molecular Biology', in S. de Chadarevian and N. Hopwood (eds.) *Models: The Third Dimension of Science*. Stanford: Stanford University Press.

Denis, M. (1909/2003). 'From Gauguin and Van Gogh to Neo-Classicism'. in C. Harrison and P. Wood (ed.) *Art in Theory 1900–2000*. Oxford: Maldon a.o., pp. 46–51.

Eco, U. (1994a). The Limits of Interpretation. Indiana: Indiana University Press.

Eco, U. (1994b). *Six Walks in the Fictional Woods*. Cambridge, MA: Harvard University Press.

Elgin, C. (1983). *With Reference to Reference*. Indianapolis: Hackett.

Elgin, C. (1996). *Considered Judgment*. Princeton, NJ: Princeton University Press.

Elgin, C. (2007). 'Understanding and the Facts'. *Philosophical Studies, 132*, pp. 33–42.

Elgin, C. (2010). 'Telling Instances', in R. Frigg and M. C. Hunter (eds.) *Beyond Mimesis and Convention: Representation in Art and Science*. Berlin and New York: Springer.

Elkins, J. (2007). *Visual Practices across the University*. Munich: Wilhelm Fink Verlag.

French, S. (2003). 'A Model-Theoretic Account of Representation (or, I Don't Know Much About Art...But I Know It Involves Isomorphism)'. *Philosophy of Science, 70*, pp. 1472–83.

French, S. (2010). 'Keeping Quiet on the Ontology of Models'. *Synthese, 172*, pp. 231–49.

Friend, S. (2007). 'Fictional Characters'. *Philosophy Compass, 2*, pp. 141–56.

Frigg, R. and J. Nguyen. (2016). 'The Fiction View of Models Reloaded'. *The Monist, 99*, pp. 225–42.

Frigg, R. and J. Nguyen. (2017a). 'Models and Representation', in L. Magnani and T. Bertolotti (eds.) *Springer Handbook of Model-Based Science*. New York: Springer.

Frigg, R and J. Nguyen. (2017b). 'Scientific Representation Is Representation As', in Hsiang-Ke Chao, J. Reiss and Chen Szu-Ting (eds.) *Philosophy of Science in Practice: Nancy Cartwright and the Nature of Scientific Reasoning*. New York: Springer, pp. 149–79.

Frigg, R. and J. Nguyen. (ms). 'The Turn of the Valve: Representing with Material Models'.

Frigg, R., J. Berkovitz and F. Kronz. (2016). 'The Ergodic Hierarchy'. *Stanford Encyclopedia of Philosophy*.

Giere, R. (2004). 'How Models Are Used to Represent Reality'. *Philosophy of Science, 71*, pp. 742–52.

Giere, R. (2010). 'An Agent-Based Conception of Models and Scientific Representation'. *Synthese*, 172, pp. 269–81.

Goodman, N. (1976). *Languages of Art*. 2nd ed. Indianapolis: Hackett.

Hartmann, S. (1995). 'Models as a Tool for Theory Construction: Some Strategies of Preliminary Physics', in W. E. Herfel, W. Krajewski, I. Niiniluoto and R. Wojcicki (eds.) *Theories and Models in Scientific Processes (Poznan Studies in the Philosophy of Science and the Humanities)*, 44. Amsterdam and Atlanta: Rodopi. pp. 49–67.

Hofbauer, J. and K. Sigmund. (1998). *Evolutionary Games and Population Dynamics*. Cambridge: Cambridge University Press.

Kennedy, A. (2012). 'A Non Representationalist View of Model Explanation'. *Studies in History and Philosophy of Science, 43*, pp. 326–32.

Knuuttila, T. (2005). *Models as Epistemic Artefacts: Toward a Non-Representationalist Account of Scientific Representation* (PhD Thesis). University of Helsinki, Helsinki.

Knuuttila, T. (2011). 'Modelling and Representing: An Artefactual Approach to Model-Based Representation'. *Studies in History and Philosophy of Science, 42*, pp. 262–71.

Kulvicki, J. (2006). 'Pictorial Representation'. *Philosophy Compass, 1*, pp. 535–46.

Leggett, D. (2013). 'Replication, Re-Placing and Naval Science in Comparative Context, C. 1868–1904'. *British Journal for the History of Science, 46*, pp. 1–21.

Lopes, D. (1996). *Understanding Pictures*. Oxford: Oxford University Press.

Mandelbrot, B. B. (1982). *The Fractal Geometry of Nature*. San Francisco, CA: W. H. Freeman & Co. Ltd.

Peschard, I. (2011). 'Making Sense of Modeling: Beyond Representation'. *European Journal for Philosophy of Science, 1*, pp. 335–52.

Putnam, H. (1981). *Reason, Truth, and History*. Cambridge: Cambridge University Press.

Salis, F. (2013). 'Fictional Entities', in J. Branquinho and R. Santos (eds.) *Online Companion to Problems in Analytic Philosophy*. Lisbon: University of Lisbon.

Salis, F. and R. Frigg. (forthcoming). 'Capturing the Scientific Imagination', in P. Godfrey-Smith and A. Levy (eds.) *The Scientific Imagination*. New York: Oxford University Press.

Smith, L. A. (2007). *Chaos. A Very Short Introduction*. Oxford: Oxford University Press.

Sterratt, D., B. Graham, A. Gilles and D. Willshaw. (2011). *Principles of Computational Modelling in Neuroscience*. Cambridge: Cambridge University Press.

Sterrett, S. G. (2002). 'Physical Models and Fundamental Laws: Using One Piece of the World to Tell About Another'. *Mind and Society, 5*, pp. 51–66.

Suárez, M. (2003). 'Scientific Representation: Against Similarity and Isomorphism'. *International Studies in the Philosophy of Science*, 17 (3), pp. 225–44.

Toon, A. (2011). 'Playing with Molecules'. *Studies in History and Philosophy of Science, 42*, pp. 580–9.

Vorms, M. (2011). 'Representing with Imaginary Models: Formats Matter'. *Studies in History and Philosophy of Science, 42*, pp. 287–95.

Webb, B. (2001). 'Can Robots Make Good Models of Biological Behaviour?' *Behavioral and Brain Sciences, 24*, pp. 1033–50.

Weisberg, M. (2013). *Simulation and Similarity: Using Models to Understand the World*. Oxford: Oxford University Press.

Werndl, C. and R. Frigg. (2015). 'Reconceptualising Equilibrium in Boltzmannian Statistical Mechanics and Characterising Its Existence'. *Studies in History and Philosophy of Modern Physics, 49*, pp. 19–31.

4 Is Captain Kirk a natural blonde?

Do X-ray crystallographers dream of electron clouds? Comparing model-based inferences in science with fiction

Ann-Sophie Barwich

1 Introduction: science and fiction

Scientific models and fiction have one noticeable feature in common. Their representational relation to the physical world is ambiguous. It is often not obvious whether certain elements in a model represent something in the world or are an artifact of a model's internal structure. Fiction, too, can mimic our world to varying degrees, as fictional worlds sometimes contain historical characters or events, such as Henry VIII or the Stonewall Riots.

When we use scientific models, however, expectations of how our inferences address the world differ from interpretations of fiction. We consider scientific models to be representations that are true about something in the world. By contrast, we regard fictions as being important records of human culture, but not as true of anything in particular. Wherein is this difference grounded, and how is it justified?

The increasing dependency of scientific research on mediated forms of observation and depiction makes this question central to philosophical interest about the characteristics of scientific representation. Laboratory conditions are hardly representative of many natural phenomena that we aim to investigate through them. That is true no matter whether we talk about the development and use of model organisms (Ankeny and Leonelli, 2011), set-ups in sensory measurement (Barwich and Chang, 2015), or studies of protein synthesis (Rheinberger, 1997). From this perspective, a large part of the philosophical debate has engaged with the deep dependency of scientific inquiry on indirect modeling practices. If modern science builds on a necessarily mediated approach, how must we understand its claim of giving us access to (the fundamentals of) reality?

In reply to this question, some philosophers have started comparing scientific representations with fiction to understand their potentially common basis as *cognitive strategies* (Suárez, 2009). Consider the case of idealizations or abstractions. These strategies provide us with scenarios of events or the behavior of entities in particular circumstances. In many cases, such as ideal gases or frictionless planes, these scenarios are not realized or realizable in the actual world. Is there something that distinguishes these scenarios from fictionalizations?

My focus in this chapter is on the special epistemic role that we assign to scientific representations as giving us indirect access to the reality of nature. A significant number of philosophical arguments have been directed at the dyadic relation between a model and its target system. Central to these arguments is an understanding of representations that addresses their structural features (e.g., isomorphism) or their conventional use as "as if scenarios." Here, I pursue an alternative tactic. Looking closely at model-building strategies, I focus on the interpretative strategies that deal with the representational limits of models. The chief question is, how do we interpret ambiguous elements in models? Moreover, how do we determine the validity of inference about information that is not explicit in a model? I suggest that the answer lies in the particular strategies that link a model to other methods in an experimental context.

In what follows, the chapter begins in Section 2 with the problem of representation in the contemporary philosophy of science. After introducing the reasons that prompted a comparison of scientific models with fiction, I argue that the problem of ambiguous inference emerges from two essential features of representations, namely their hybridity and incompleteness. To distinguish between fictional and non-fictional elements in scientific models, my proposal is to look at the integrative strategies that link a particular model to other methods in an ongoing research context. To exemplify this idea, I examine protein-modeling through X-ray crystallography as a pivotal method in biochemistry. As many readers from the humanities and the arts may not be familiar with this method, Section 3 introduces the procedure in greater detail. My reason for this is to allow the reader to judge whether she considers my concluding analysis of the underlying fictionalization strategies in Section 4 as adequate.

2 Context and argument: the problem of representation

What are the origins of the philosophical debate about whether a scientific representation depicts reality accurately? Besides, what epistemological concerns suggest a comparison of scientific models with fiction? These questions have an inevitable historical dimension, so three main factors must be pointed out briefly. First, there is the legacy of radical theory changes, especially in nineteenth-century chemistry and twentieth-century physics. These changes shattered the epistemic confidence in our scientific methods and models. How can we be sure that our current theories and representations are more truthful than those left to rot in the graveyard of scientific history? Will our current concepts and models fail us, too? This issue is known as the pessimistic meta-induction in the philosophy of science, and if history is any indicator, a cautious attitude towards proper candidates for truth is advisable.

Second, the rise of studies that recognized science as an essentially social and historical endeavor further substantiated this caution. Social studies of science, especially over the second half of the twentieth century, demonstrated the contingency of factors that shape scientific advancement (Longino, 1990; Barnes, Bloor, and Henry, 1996). Third, when contemporary philosophy picked up on these developments, it turned away from broad generalizations about the nature of science. Instead, scholars directed their attention to the cognitive and experimental practices that underlie scientific research, for instance, in the construction of scientific concepts and representations such as models (Morgan and Morrison, 1999; Nersessian, 2010).

One line of investigation emerging from these advancements is the recent interest in the problem of representation (Knuuttila, 2005). As artifacts of human activity, how can scientific representations provide us with access to the world? How do we use these representations to gain knowledge about real things? Moreover, how must we understand cases of representational failure? On what basis are our model-based inferences justified? These questions have occupied many philosophers over the past decade. Their arguments revealed that scientific representations, as cognitive tools, often rely on fictionalization strategies or features that are shared by forms of fiction.

The meaning of fiction in this context is that of mimicry and distortion (Frigg, 2010a). Fiction as mimicry refers to representations that are designed to resemble real phenomena without truly referring to them, or without claiming to be a proper or truthful representation of these phenomena. In comparison, fictions as distortions describe alterations that present an (intentionally or unintentionally) alienated or converted image of a phenomenon. A differentiation of these two meanings, mimicry and distortion, is not necessarily clear-cut. Satire is a wonderful example of the thin conventional – and legal – line that holds between the mimicry of something (or somebody) in parallel with the explicit distortion of its features. Overall, this understanding of fiction centers on two fundamental philosophical topics: the reference of representations (or its suspension) on the one hand, and the truthfulness or accuracy of representations on the other.

Corresponding to this idea of fictionality, some philosophers of science encountered similar issues in the analysis of representations in science (for a collection of essays, see Suárez, 2009). Almost all scientific models build on various forms of distortions, abstractions, idealizations, imaginative scenarios, metaphorical comparisons, analogies, and so on (Hesse, 1966; Holyoak and Thagard; 1996; van Fraassen, 2010). The influential "billiard ball" model of the atom by Dalton, the rise of the ball-and-stick model of chemical substances by August Wilhelm von Hofmann, or the Homo Economicus in economic theories are just some of the many prominent examples. Furthermore, many scientific entities were believed to be true initially, but turned out not to exist. Consider the abandoned scientific concepts of the ether, pneuma, phlogiston, or the idea of a *spiritus rector.* Nevertheless, these entities yielded experimental results and temporary insights into some aspects of the world, as it has been argued convincingly for the case of phlogiston (Chang, 2012).

However, the comparison of science and fiction is not immediately intuitive. It does not come as a great surprise to scientists that we should not take scientific representations as literal depictions of the world. Neither does it sound astonishing that science rests on a graveyard of theoretical entities and proceeds through self-correction by its very nature. Science continuously changes and expands the scope of our knowledge by pointing at what we do not yet know. From this view, it is an infinite business (Deutsch, 2011). To be sure, these constitute central philosophical insights into the nature of science that have been gained over the past century. So, what does the comparison of science with fiction tell us from a contemporary perspective?

It stands to reason that science and fiction are not the same things. We do ascribe scientific representations a different epistemic status. They tell us something about what is real. However, the appropriate grounds for this claim are not always clear. In response, there are several angles from which we can approach the special status of scientific representations.

One way is to compare the structural features of scientific and fictional representations. What constitutes the resemblance or the similarity between a representation and its target system? Also, what are the functions of distortions in this context? Some philosophical arguments defined representation as a straightforward mapping connection that relates a model structure to a physical structure. An example is isomorphism: "A Model M is a structure; and M represents a target system T if T is structurally isomorphic to M" (Frigg, 2002). By contrast, fiction does not seem to rest on such a correspondence. On this account, structure as a representational criterion is based on a dyadic comparison between models and target systems.

Meanwhile, several arguments have shown that such a comparison does not lead to a clear demarcation of non-fictional from fictional representations. Criteria that seek to define the capacity of models to represent a target system through purely structural criteria must fail eventually (Goodman, 1969; Frigg, 2002). My view on this issue is that criteria of "similarity" and "structure" (whatever these may be) are indeed insufficient. Though, I cannot help thinking that the use of structure in these arguments presents a straw man.

Noticeably, this approach runs into trouble if we consider models as conglomerates and as consisting of different model ingredients (Boumans, 1999). From this perspective, the representational function of structures is contingent on the context of model-building. This point of view led to an emphasis on the epistemic context in which scientific representations are used (Knuuttila and Voutilainen, 2003). Here, the representational capacity of models and other scientific representations, such as diagrams or algorithms, is defined primarily by their epistemic function instead of their structural correspondence. Thus representations are understood to act as vehicles for reasoning and for making indirect inferences (Suárez, 2004). This means that representations are part of an imaginative act of "make believe." They present us with "as if scenarios" that give us "fictional" as in model-dependent truths (Walton, 1990; Frigg, 2010b; Toon, 2012). This idea strongly relies on a conventionalist understanding of an institutional or collective agreement about the use of representations (Searle, 2010).

However, such reference to the use and conventions of an epistemic context seems to beg the question. It presupposes that we already know how to use (parts of) a representation adequately (i.e., as denoting or non-denoting). In fact, this account even dodges the real issue of a comparison between fiction and non-fiction as long as it avoids an answer to the question about the particular character of scientific representations. In essence, on what grounds is something referred to as a non-fictional representation? How can we find out whether an inference is only a "fictional truth," as in an essentially model-dependent truth, or whether it accounts for something model-independent but real?

In reply to this question, I propose an alternative tactic. Representations are neither judged by structural features in a dyadic comparison nor seen as make-believe scenarios. Rather, my focus is on the particular strategies that we employ in the interpretation of representations, and how these strategies help us to distinguish between what may be a fictional (model-dependent) and non-fictional (representational) model-component. My argument consists of two claims about the ontology of representations, fictional as well as non-fictional.[1] These are as following: First, most representations, fictional and non-fictional, are hybrids. Second, every individual representation is incomplete.

Beginning with the first claim, hybridity means that most representations contain denoting as well as non-denoting elements (for a more detailed argument see Barwich, 2013). For example, Bulgakov's fictional novel *The Master and Margarita* tells us a story about the devil having a ball in the Moscow of the 1930s. Theological disputes aside, the devil is not what most people would consider a real person today. Nonetheless, there was such a place as the Moscow of the 1930s. For a complementary example, consider the ball-and-stick models of chemical substances as a scientific representation. These models account for the basic spatial organization of a molecule, but in a highly idealized sense. In light of such mixed characteristics, it is often not intuitive which individual elements in a representation denote and which do not. Moreover, do these elements denote independently of the overall epistemic status of the representation in which they are contained (i.e., does it matter whether these elements are part of a novel or a scientific model)? For example, does the fictional Napoleon in *War and Peace* refer to the real historical figure of Napoleon? More so, what if we encounter a fictional Napoleon that has little in common with the descriptions of the historical figure? What about the pig Napoleon in George Orwell's *Animal Farm*? Besides, are non-existent entities that are part of working scientific models (such as silogens) fictions?[2] Furthermore, can fictional elements be accidentally true? Consider the possibility that we find a real person that matches every description of a fictional character's biography – without the author's (and perhaps even that individual's) knowledge. Is the fictional character suddenly a true description of this real person? There have been different answers to this problem, so it seems that the response depends on one's personal philosophical predilections (Ryle, 1933; Danneberg, 2006). Any position on this issue appears to presuppose a specific understanding and definition of what the notions of fictional and non-fictional entail. A clear demarcation between fiction and scientific representations based on their denoting and non-denoting elements is inadequate.

Alternatively, we must look at representational functions in context. In context means analyzing what the individual components contribute to the containing representation in question. This interpretative setting serves as the basis for inferences about these elements (Danneberg, 2006). On this account, neither the Napoleon in Tolstoi's *War and Peace* nor the pig Napoleon in Orwell's *Animal Farm* denotes the real Napoleon. Instead, these two figures refer to our knowledge about the real Napoleon without being used to argue or certify any claim about Napoleon as a historical person. Therefore, these fictional Napoleons constitute an image of a denoting element, meaning they are not used to denote but refer to other items that are. The function of Napoleon as an element in *War and Peace* is determined by its role in the fictional world-story, not by its epistemic connection to historical sources (Barwich, 2013).

My second claim states that each form of representation is incomplete. Incomplete means representations are limited and selective in their descriptions concerning real world properties. For example, a map only contains specific elements depending on its purpose (e.g., transportation maps give you information about the underground and bus system, but not about altitude). Notably, advocates of scientific pluralism present a similar argument when they concern scientific models. Models are necessarily limited in their content and scope. Their boundaries resonate with their epistemic objective: Are we aiming for realism, generality, or precision when using a model (Levins, 1966)? This methodological argument for pluralism derives from the ontological complexity of the physical world. To capture the multi-leveled, overlapping, and contingent features of nature, especially of biological entities and processes, we need a mosaic of models (Mitchell, 2003; Wimsatt, 2007).

Now, this has further consequences for our interpretation of such models. We sometimes look for information that is not explicit in a representation (for a more detailed argument see Barwich, 2014). Sometimes this information is implicit. For example, reading a novel about two people falling in love, you possibly assume that these two individuals have a heart, a liver, and at least one functioning kidney, even if the author fails to mention this. It is because we know that people usually do not survive without these organs. We can make such explicit inferences even in fiction because any representation is somewhat "parasitic" on our common knowledge about the world and the use of language concepts (Searle, 1975; Eco, 1994). Therefore, unless stated explicitly, any word or concept in fictional discourse has the same meaning and implications as it has in non-fictional discourse.

What if we want to know about something that is not inferable in such an implicit sense? Some questions lead us to the limits of interpretation if they are not answerable by inferences based on the representation in question. For example, ask yourself, *how many children had Lady Macbeth?* There is no plain answer to this issue. Shakespeare's Lady Macbeth may not be of the nurturing kind, but she could have had children, even though none are mentioned explicitly in the play (Knights, 1933).[3] Likewise, we know that Captain James T. Kirk from the starship *Enterprise* is blonde. However, is he a natural blonde?

The vanity of his character does not exclude the possibility that he colors his hair regularly. We will never know for sure as long as no conclusive information is provided. The same issue goes for the interpretation of scientific models. The chemical formulas of Berzelius, for example, give information about proportions but not about mechanical features of atoms, such as their size or shape (Klein, 2001). Nonetheless, in an ongoing experimental pursuit, such missing information is not left aside. It is addressed by amending the model or, alternatively, by looking for an answer through alternative models in the same research context. Identifying and clarifying such limits of models spurs further scientific inquiry.

Persistent failure to provide a better account of such missing information in a model can also lead to serious doubt about the model's referential grounds. Consider the case of phlogiston. Central to its ontology was the question of whether phlogiston is weightless, and the continuing inconsistencies in answering this issue were a reason for abandoning the concept (Kim, 2008; Barwich, 2013).

The upshot is that identifying and testing the limits of scientific representations is a useful way to determine whether a particular inference is only a model-dependent, fictional truth. Overall, this section outlined two central characteristics of representations, namely hybridity and incompleteness, which further allow us to compare and distinguish fictional and non-fictional representations. The remainder of the chapter now elaborates on how we can use these two characteristics to make decisions about ambiguous elements in scientific modeling. For this, I now look at how scientists make sense of indeterminate inferences when using X-ray crystallography for protein models. Indeed, X-ray crystallography is an excellent example to analyze the problem of representation, as we will see that crystallographers face several of the representational issues mentioned above.

3 The case of X-ray crystallography

X-ray crystallography is a principal method in biochemistry. It serves to determine the molecular structure of macromolecules such as proteins, and it works like this: Protein materials are prepared in specific detergents so that they form neat crystalline structures. These crystals are mounted on a goniometer (an instrument that allows the rotation of the inserted object). When placed in the goniometer, the crystals are shot with beams of X-rays. These X-rays scatter on the crystal surface and form a diffraction pattern, which is collected on an image plane or X-ray film. This diffraction data accounts for the electron density in the crystal structure. It serves as the basis to infer atom positions and, consequently, the molecular structure of the crystal (Drenth, 2007; Serdyuk, Zaccai, and Zaccai, 2007; Smyth and Martin, 2000).

Successful applications of this procedure are not without trickery and difficulties, of course. There is no simple mapping of the models onto the raw materials. Rather, instrumental access to macromolecules is facilitated through several steps of manipulating the materials to fit the model requirements. The foundation of protein models in X-ray crystallography is not so much a reconstruction but the very production of certain structures. The success of inferences based on this method is contingent on the multiple steps for bringing the materials in correspondence with the requirements of the model procedure. Each of these steps carries its ambiguities and modeling problems.

3.1 Distortions, or: making the materials fit the method

The essential material precondition for X-ray crystallography is symmetry. So, why do we need symmetric crystals in the first place? Creating symmetric structures from organic matter is a difficult issue, but essential for successful data collection.

To make crystals, you start out with proteins in a solution of high concentration. You purify them and slowly remove the water by putting the materials in a specific detergent. However, the crystallization of proteins can be a daunting task, and it often involves a laborious trial-and-error procedure. "The magnitude of the problem can be understood when one considers the variables: the choice of precipitant, its concentration, the buffer, its pH, the protein concentration, the temperature, the crystallization technique, and the possible inclusion of additives" (Smyth and Martin, 2000: 8).

Some macromolecules form nice, symmetric crystalline structures. Others remain stuck in an unsymmetrical and flat formation. A particularly salient example for this are transmembrane proteins, such as the superfamily of G-protein-coupled receptors. G-protein-coupled receptors, or GPCRs, constitute the largest protein gene family in the mammalian genome, and are involved in several fundamental biological processes such as vision, olfaction, the regulation of immune responses, and the detection of neurotransmitters. However, their instability prevents these transmembrane proteins from building regular crystalline structures easily. The first success in getting the structure of an active ternary complex of a GPCR through X-ray crystallography happened only a few years ago (Snogerup-Linse, 2012; Kobilka, 2013).[4] Any attempt to crystallize proteins from the largest member of this protein superfamily, the olfactory receptors, has been unsuccessful to date (Crasto, 2009; Barwich, 2016).

These difficulties aside, regular applications of X-ray crystallography require three-dimensional symmetric crystals. Symmetry is indispensable for combining the series of diffraction images that are collected while the crystal is rotated. Symmetry allows applying mathematical interpretations to these diffraction patterns and to turn them into electron-density maps. The reason for this is because crystals facilitate the determination of "unit cells," those parts of the crystal that form its smallest repeating units. These unit cells act as a metric index to determine the dimensions of the overall crystal structure. They further relate the series of individual diffraction images to each other: "Computer programs for autoindexing do this by calculating a prediction of what the diffraction image will look like from the cell dimensions and orientation, then attempting to fit the real image with the predicted one" (Smyth and Martin, 2000: 12).

This is why you need symmetric crystals for the method to work. Data from flat or unsymmetrical crystals are incomplete, distorted, and impossible to accommodate with the mathematical tools (Smyth and Martin, 2000). Distorted and indeterminate data pose significant limits for legitimate inferences to the molecular structure and constitution of macromolecules, such as proteins. However, even obtaining "good enough" diffraction data has its difficulties and limits. As mentioned earlier, some proteins are harder to crystallize than others, such as transmembrane receptors. Furthermore, proteins, when treated with X-rays, disintegrate quickly and, as a result, the collected diffraction data can be incomplete or insufficient. Concerning the gradual disintegration of proteins during the procedure, the first diffraction image is usually of the best quality.

This issue of indeterminate data is more than a matter of temporary technological concern. It marks the methodological dependence of experimental research on the available instrumental tools and the structural aspects of the material to which they are responsive. Successful inferences to the molecular dimension of proteins do not ground in an essential trait or structure of the raw materials per se. On the contrary, proteins are irregular and dynamic when untreated. Instead, methods such as X-ray crystallography rely on artificially-produced features, such as symmetry. That said, the production of symmetric crystals is not the only methodological requirement for successful model-building through X-ray crystallography.

3.2 Incomplete information in model-based inferences

The next basic step is the analysis of diffraction data and their transformation into electron-density maps. What we see in recordings of diffraction patterns is a distribution of electron density. The concentrated rings of spots are the result of diffracted X-rays that are emitted from the crystal and collected on an imaging plane. These spots indicate electron clouds. Now, how do we get to the molecular structure of proteins from a measure of electron density? Meaning, how do we infer atom positions from these electron spots?

This is a matter of data-processing. The analysis of diffraction patterns rests on the distinction between meaningful diffraction data and mere background noise. This distinction is crucial for further calculations and to transform the diffraction data into readable electron density maps. These density maps then allow us to infer atom positions and the molecular organization of the protein. Notably, in the early applications of X-ray crystallography, this was the very problem. What is the noise to data ratio? Moreover, what do we see when we look at the diffraction data?

X-ray beams come in waves, and waves have different phases. However, we can only record the overall intensity of beams on the image planes. Thus, the recordings do not tell us anything about the particular phase the waves are in when hitting the plane. Are these waves in sync (in phase) or not (out of phase)? To be sure, different phases result in different intensities of the recorded spots. However, to understand what these diffraction spots represent, and to infer atom positions, we need to know in which phase the X-ray beams are when they hit the plane. How does one get access to this information? The mere diffraction patterns do not allow for inferences to the wave phases, and the issue is known as the "phase problem."

Solving the phase problem was an ingenious piece of work by the molecular biologist Max Perutz at Cambridge, 1962 Nobel Laureate in Chemistry (Perutz, 1942; 1962). Perutz wanted to solve the molecular structure of hemoglobin, a complex protein. He saw himself confronted with the missing phase information, resulting in a lack of representational stability and a strong ambiguity in data recordings. The solution to this query he obtained by adopting a method previously used by one of his colleagues, J. M. Robertson at Glasgow University: isomorphous replacement (Pietzsch, 2016).

Isomorphous replacement works by soaking the protein in a heavy atom solution, such as mercury or platinum. The protein incorporates one or more of these heavy atoms, but this does not change its spatial configuration. Nonetheless, the incorporation of heavy atoms modifies the diffraction patterns, since the "heavier" protein now contains more electrons. Therefore, we end up with two sets of data: the diffraction patterns from the original protein and the patterns of the protein containing heavy atoms. Comparing these data sets, any differences in recorded intensities are due in large part to the presence of heavier atoms. From this we can derive the missing phase information. The subtle differences in intensities between the data sets allow us to find the position of the heavier atoms. These atoms then act as reference points to infer the phases (Smyth and Martin, 2000; Cowtan, 2003; Pietzsch, 2016).

The point that the resolution of the phase problem illustrates is that the interpretation of structural correspondence is a product of material manipulations. Therefore, inferences derived from these structures must be judged carefully against the model-building requirements. What we see through such a detailed perspective on the process of model-building is again that the structural correspondence between a model and the materials reflects the methodological conditions of the procedure, and not necessarily essential, raw features of the materials. To be sure, presumptions about structural features are an integral part of modeling. Nonetheless, they cannot provide independent criteria for the evaluation of model-based inferences and the referential capacity of a scientific representation.

The third step is building the final protein model. For this, the quality of the electron-density map is crucial to determine the arrangement of molecular units for a three-dimensional model. Electron maps provide the basic outline wherein the protein structures are built. The higher the resolution of these maps, the less ambiguous is the identification of the relevant molecular units and subunits. The derived image presents us with a three-dimensional model of the protein structure. Lastly, the outputs are formatted and placed in the protein data bank (PDB).[5] An integral part of this concluding step is the continuous refinement of the data analysis. Specifically designed molecular visualization programs facilitate the interpretation of structural features of the molecule, such as bond lengths and angles (Smyth and Martin, 2000).

4 Integrative pluralism as a representational requirement in scientific practice

At this point, let us come back to the general issue of representation and the epistemic role we assign to scientific representations. In examining the details of X-ray crystallography, I deliberately avoided discussing how the problems in this procedure resonate with the debate about fictionalization strategies in scientific representation. My reasons for this was because I wanted to focus in more detail on how this method works from the practitioners' perspective before connecting its modeling steps to the philosophical discussion at hand.

To bring the different modeling steps together for further analysis, so far we have seen that multiple factors play a major role when building a protein model through X-ray crystallography. These factors concern the experimental manageability of the research materials, the generation of a sufficient range of data, the availability of appropriate methods to translate the diffraction patterns into electron-density maps, and also the introduction of data-processing techniques, such as molecular visualization programs. All these factors are involved in a series of material and conceptual operations that shape and result in the final receptor model. In brief, these sequential steps comprise:

- first, the material transformation of flexible proteins into stable and rigid crystal structures (fulfilling the requirement of symmetry);
- second, the material inscription from crystallized protein structure to diffraction data (relying on Bragg's model, or Bragg's law);[6]
- third, the translation of diffraction data into electron density maps through the Fourier transform);[7]
- fourth, the subsequent inferences from the electron density map to a three-dimensional model of the protein (using computational programs to calculate and visualize the positions and relations of atoms as inferred from the electron clouds).[8]

The detailed dissection of modeling steps in X-ray crystallography revealed a chain of mediation between the raw materials and the final model. This chain made explicit the several requirements by which we generate "structures" as a meaningful model ingredient. Although the successive modeling steps do not follow logically from each other, the manifestation of these steps is informed by, and grounded in, the inferences and structural correspondences established through their preceding ones. On this account, an analysis of scientific representations based on dyadic and abstracted notions such as "structural similarity" or "resemblance" between a model and its target system must look too simplistic to the practitioner to be of practical utility. To her, it matters more to suggest a heuristic perspective on how to deal with the ambiguous components in a modeling procedure, such as the phase problem. For this, a more detailed understanding of the method is necessary.

Every model has a history that determines its construction and use and, in turn, its potential and limits for inference to the properties of the investigated materials. The models we derive from scientific methods are, therefore, not freely floating objects that correspond to the world through some depiction of abstracted features. Instead, these abstracted features are a representation of the steps that link specific structural assumptions to the materials in a series of manipulations. From this view, any evaluation of a model and the inferences drawn from it must be judged against the background knowledge on which the modeling procedure is based. In the case of crystallography, to assess how the derived protein structures do (or do not) account for protein properties and behavior, a mere look at the finalized end product is insufficient.

The fictionalization strategies we encountered here included: (1) distortions (forcing proteins into symmetric crystals); (2) idealizations (predictions of crystal structure through autoindexing of unit cells to align diffraction planes); (3) indeterminate data (disintegrating crystals); and (4) the underdetermination of models by data (recordings of electron spots as part of the phase problem). Indeed, it appears miraculous that this procedure could become one of the most reliable and standard methods in structural biology and biochemistry at all!

That said, these issues demonstrate the relevance of my two earlier hypotheses, namely the hybridity and the incompleteness of representations. Regarding the first hypothesis, these fictionalization strategies embody the hybridity of crystallography models where certain elements do not strictly denote features of proteins but reflect the requirements of the modeling method. Concerning the second hypothesis, the incompleteness of representations, we may ask: How are the ambiguous model elements dealt with in X-ray crystallography? Accordingly, how do the practitioners answer those questions about the materials that exceed the available model structures? The kinds of issues that exceed a method usually arise from the particular purpose a method is designed to serve.

The target of protein models, constituting the Holy Grail of modern biochemistry, is to solve the relation between a protein's structure and its function. Thus far, however, drug target studies and approaches to rational drug design operate on a labor-intensive trial and error basis (Drews, 2000). A pivotal reason for the lack of principles after which to model structure-function relations in proteins is the problem of protein-folding. After decades of elaborating on protein structures, it became apparent that there is no straightforward way to make predictions from the amino acid motifs to the three-dimensional configuration of proteins.[9] Proteins often fold irregularly. However, the folding process fundamentally determines protein function. As a result, model-based inferences to ligand-binding in proteins require three-dimensional snapshots of the proteins. This issue also explains why X-ray crystallographers are still in business (Mitchell and Gronenborn, 2015).

Nevertheless, the irregularity between protein structure and function not only explains the continuing demand for crystallography. It also presents the limits of this procedure. A major concern for biologists is precisely what the protein models from X-ray crystallography do not show and the questions they cannot answer.

Proteins are extremely specific in behavior and function. Their functioning, such as in molecular recognition, is based on a highly dynamic mechanism, involving several steps of conformational changes of proteins. Modern biology has therefore abandoned the lock-and-key metaphor. Instead, biologists refer to models of "induced" (Koshland, 1995) and "selected fit" (Monod, Wyman, and Changeux, 1965; Changeux, 2013). Here, proteins are not rigid but dynamic entities. They undergo multiple changes in their conformation. Several questions arise from such models of conformational changes, such as: Are these changes induced by the ligand, or do these changes take place in the absence of a ligand? (For more details on these mechanisms see Barwich, 2016.)

X-ray crystallography is unsuitable for determining the sequences of flexible conformational changes in protein-binding. It is also unfit to make inferences about the specific causal role of the ligand in the recognition mechanism. Yet, these are two essential issues for understanding protein behavior. From this view, the strength of this method is also the source of its ambiguity: How can we make inferences about protein behavior from the rigid representations of protein structure?

Despite its overall success, concern about the limits of X-ray crystallography has always been vocal. A source of such concern often goes back to the inevitable incompleteness of representations. We have seen this to be the case for inferences about the binding behavior of proteins. Additionally, the characteristic of incompleteness is inevitably part of the model-building procedure, not only of the final product. For example, the big issue in the earlier applications of this method was the ambiguity of what the image planes show. When Perutz presented his diffraction patterns, it was Francis Crick, his student, who ambushed him severely in a public talk regarding the ambiguity of this data in light of the unresolved phase problem (Pietzsch, 2016). The issue was resolved after an additional procedure was implemented into the model. The data from isomorphic replacement resonated precisely with the structural requirements of the method, and this correspondence then allowed a determination of the unknown phases. To be sure, once the phase problem was solved, other limits of the procedure were rendered visible.

Scientists rarely judge the interpretation of a model in isolation. Additional procedures are used in parallel with a model to determine which parts of its inferences can be corroborated and complemented by other means or may constitute a potential artifact. Of particular importance is a satisfactory independence of some modeling parameters in the corroborating procedure. Of course, the issue then is to ascertain that the different procedures account for the same phenomenon and are coordinated with each other (Barwich and Chang, 2015).

For example, complementary to X-ray crystallography, another way to build protein models is nuclear magnetic resonance (NMR) spectroscopy. One of the limits of crystallography is the requirement of isolating the protein from its cellular environment and studying it through a crystalline state. In NMR, proteins are used in a soluble state. "Furthermore, X-ray crystallography and NMR target different atomic features of a protein: X-ray crystallography relies on the scattering of X-rays by the electrons, while in NMR, interactions of nuclear spins with a magnetic field are explored (Mitchell and Gronenborn, 2015: 15)." In comparison, each method results in mediated representations of proteins outside their cellular environment. Still, both approaches count as reliable representational sources for inferences about protein behavior in vitro.

Overall, the difficulty of indeterminate inferences – meaning inferences that may not represent a physical entity but constitute a model artifact – are addressed through a comparison of data through different procedures. Such comparisons can be an indicator of inaccuracies and a reflection of a method's biases. To correct these biases requires a deeper understanding of the structural requirements that fit the materials to the method in the model-building procedure.

In closing, we must understand inferences to information that is not explicitly given in the model as means to determine how this model is linked to other methods in an experimental context. In contrast to fiction, interpretations of ambiguous elements and incomplete representations are how science proceeds. From this perspective, the epistemic distinction between science and fiction is not one inherent in a representational structure but in our way of dealing with such limits. Therefore, the integration of a scientific model with other methods ensures that its inferences are testable as being non-fictional (i.e., representative of a phenomenon) or as fictional (i.e., model-dependent and an artifact of the procedure). As science is ongoing, a scientific representation cannot be final. Therefore, its representational function and limits must be probed through its integration in an operating and ongoing model context.

Notes

1 By representations, I mean all forms of public description and depiction, whether these are linguistic, algebraic, symbolic, pictorial, and so on.

2 Silogens are hypothetical entities in nanomechanical models. They are used to calculate silicon fractures and combine theoretical assumptions from quantum mechanics and classical molecular dynamics. Silogens do not refer to real atoms. Instead, they are algorithmic combinations of the properties of two different entities in the modeling procedure, namely silicon and hydrogen (Winsberg, 2009). Notably, my interpretation of the epistemic status of silogens differs from Winsberg's (Barwich, 2014).

3 In a famous essay, entitled 'How Many Children Had Lady Macbeth?', the Shakespeare scholar L. C. Knights (1933) asked about the inferential limits in fictional works. This essay argued against the tendency of excessive over-interpretations in literature studies at the time.
4 Brian Kobilka received the 2012 Nobel Prize in Chemistry for this achievement. He shared this nomination with his former mentor Robert Lefkowitz, who received the award for his crucial work on the ß-adrenergic receptors and the general workings of GPCRs (Snogerup-Linse, 2012; Clark, 2013).
5 Website of the PDB: http://www.wwpdb.org
6 Cambridge physicist William Lawrence Bragg and his father William Henry Bragg, professor of mathematics at the University of Adelaide, determined the angles in which X-rays scattered from a crystal lattice. Together they were awarded the 1915 Nobel Prize in Physics. Lawrence Bragg was only 25 at the time and has been the youngest Nobel Prize winner in Physics to date (Nobel Media, 2016).
7 The Fourier transform allows calculating the frequencies that make up a signal (here, the X-ray beam). The Fourier transform is named after the nineteenth-century French mathematician and physicist Joseph Fourier.
8 Examples for such molecular visualization programs are RasMOL and MOLMOL.
9 When we speak of protein structure, we must differentiate between four levels: primary structure (amino acid sequences), secondary structure (regular subunits of a protein such as helix domains), and tertiary structure (three-dimensional protein folding), and the quaternary structure (three-dimensional structure of multi-subunit proteins).

Bibliography

Ankeny, R. A. and S. Leonelli. (2011). 'What's So Special about Model Organisms?'. *Studies in History and Philosophy of Science Part A*, 42 (2): 313–23.

Barnes, B., D. Bloor and J. Henry. (1996). *Scientific Knowledge: A Sociological Analysis*. Chicago, IL: University of Chicago Press.

Barwich, A.-S. (2013). 'Science and Fiction: Analysing the Concept of Fiction in Science and Its Limits'. *Journal for General Philosophy of Science*, 44 (2):357–73.

Barwich, A.-S. (2014). 'Fiction in Science? Exploring the Reality of Theoretical Entities', in G. Bonino, G. Jesson and J. Cumpa (eds.) *Defending Realism: Ontological and Epistemological Investigations*. Boston, Berlin and Munich: de Gruyter: 291–309.

Barwich, A.-S. (2016). 'What Is So Special about Smell? Olfaction as a Model System in Neurobiology'. *Postgraduate Medical Journal*, 92: 27–33.

Barwich, A.-S. and H. Chang. (2015). 'Sensory Measurements: Coordination and Standardization'. *Biological Theory*, 10 (3): 200–211.

Bassophile 2007. 'Eden.png.' *Wikipedia*. Accessed 26 June. https://en.wikipedia.org/wiki/File:Eden.png.

Benjah-bmm27. (2007). 'Wurtzite-unit-cell-3D-balls.png.' *Wikipedia*. Accessed 26 June. https://en.wikipedia.org/wiki/File:Wurtzite-unit-cell-3D-balls.png.

Boumans, M. (1999). 'Built-in Justification', in M. Morgan and M. Morrison (eds.) *Models as Mediators: Perspectives on Natural and Social Science*. Cambridge: Cambridge University Press.

Chang, H. (2012). *Is Water H2O?: Evidence, Realism and Pluralism*, Volume 293, New York: Springer.

Changeux, J-P. (2013). '50 years of Allosteric Interactions: The Twists and Turns of the Models'. *Nature Reviews Molecular Cell Biology*, 14: 819–29.

Clark, R. B. (2013). 'Profile of Brian K. Kobilka and Robert J. Lefkowitz, 2012 Nobel Laureates in Chemistry'. *Proceedings of the National Academy of Sciences*, 110 (14): 5274–5.

Cowtan, K. (2003). 'Phase Problem in X-ray Crystallography, and Its Solution'. *Encyclopedia of Life Sciences*. Chichester: John Wiley & Sons Ltd.

Crasto, C. J. (2009). 'Computational Biology of Olfactory Receptors'. *Curr Bioinform*, 4 (1): 8–15.

Dahl, J. (2006). 'X-ray Diffraction Pattern 3clpro.jpg'. *Wikipedia*. Accessed 23 June. https://en.wikipedia.org/wiki/File:X-ray_diffraction_pattern_3clpro.jpg.

Danneberg, L. (2006). 'Weder Tränen noch Logik. Über die Zugänglichkeit fiktionaler Welten', in U. Klein, K. Mellmann and S. Metzger (eds.) *Heuristiken der Literaturwissenschaft. Einladung zu disziplinexternen Perspektiven auf Literatur*. Paderborn: Mentis: 35–83.

Deutsch, D. (2011). *The Beginning of Infinity: Explanations that Transform the World*. London: Penguin UK.

Drenth, J. (2007). *Principles of Protein X-Ray Crystallography*. New York: Springer Science & Business Media.

Drews, J. (2000). 'Drug Discovery: A Historical Perspective'. *Science*, 287 (5460): 1960–4.

Eco, U. (1994). *Six Walks in the Fictional Woods*. Cambridge, MA: Harvard University Press.

Frigg, R. (2002). 'Models and Representation. Why Structures Are Not Enough'. *Measurement in Physics and Economics Project Discussion Paper Series*, London School of Economics. Accessed 31 February. http://www.romanfrigg.org/writings/Models_and_Representation.pdf.

Frigg, R. (2010a). 'Fiction in Science', in J. Woods (ed.) *Fiction and Models: New Essays*. 247–87. Munich: Philosophia Verlag.

van Fraassen, B. C. (2010). 'Scientific Representation: Paradoxes of Perspective'. *Analysis*, 70 (3): 511–14.

Frigg, R. (2010b). 'Models and Fiction'. *Synthese*, 172 (2): 251–68.

Goodman, N. (1968). *Languages of Art: An Approach to a Theory of Symbols*, 2nd edition. Indianapolis: The Bobbs-Merrill Company.

Haade, Wjh31, and Quibik. (2010). 'Interference of two waves.svg'. *Wikipedia*. Accessed 26 June. https://en.wikipedia.org/wiki/File:Interference_of_two_waves.svg.

Hesse, M. B. (1966). *Models and Analogies in Science*, Volume 36. Notre Dame: University of Notre Dame Press.

Holyoak, K. J. and P. Thagard. (1996). *Mental Leaps: Analogy in Creative Thought*. Cambridge, MA: MIT Press.

Kim, M.G. (2008). 'The "Instrumental" Reality of Phlogiston'. *Hyle*, 14 (1): 27–51.

Klein, U. (2001). 'Berzelian Formulas as Paper Tools in Early Nineteenth-Century Chemistry'. *Foundations of Chemistry*, 3 (1): 7–32.

Knights, L. C. (1933). *How Many Children Had Lady Macbeth?: An Essay in the Theory and Practice of Shakespeare Criticism*. Cambridge: The Minority Press.

Knuuttila, T. (2005). 'Models, Representation, and Mediation.' *Philosophy of Science*, 72 (5): 1260–71.

Knuuttila, T., and A. Voutilainen. (2003). 'A Parser as an Epistemic Artifact: A Material View on Models'. *Philosophy of Science*, 70 (5): 1484–95.

Kobilka, B. (2013). 'The Structural Basis of G-Protein-Coupled Receptor Signaling (Nobel Lecture)'. *Angewandte Chemie International Edition*, 52 (25): 6380–8.

Koshland, D. E. (1995). 'The Key-Lock Theory and the Induced Fit Theory'. *Angewandte Chemie International Edition in English*, 33 (23–24): 2375–8.

Levins, R. (1966). 'The Strategy of Model Building in Population Biology'. *American Scientist*, 54 (4): 421–31.

Longino, H. E. (1990). *Science as Social Knowledge: Values and Objectivity in Scientific Inquiry.* Princeton: Princeton University Press.

Mayer, D. (2007). 'Hexagonal lattice.svg'. *Wikipedia*. Accessed 26 June. https://en.wikipedia.org/wiki/File:Hexagonal_lattice.svg.

Mitchell, S. D. (2003). *Biological Complexity and Integrative Pluralism*. Cambridge: Cambridge University Press.

Mitchell, S. D. and A. M. Gronenborn. (2015). 'After Fifty Years, Why Are Protein X-ray Crystallographers Still in Business?' *The British Journal for the Philosophy of Science*. http://doi.org/10.1093/bjps/axv051.

Monod, J., W. Jeffries and J-P. Changeux. (1965). 'On the Nature of Allosteric Transitions: A Plausible Model'. *Journal of Molecular Biology*, 12 (1): 88–118.

Morgan, M., and M. Morrison (eds.). (1999). *Models as Mediators: Perspectives on Natural and Social Science*. Cambridge: Cambridge University Press.

Nersessian, N. (2010). *Creating Scientific Concepts*. Cambridge, MA: MIT Press.

Nobel Media. (2016). 'Facts on the Nobel Prizes in Physics'. Nobelprize.org. Accessed 26 June, 2016. http://www.nobelprize.org/nobel_prizes/facts/physics/.

Perutz, M. F. (1942). 'X-ray Analysis of Haemoglobin'. *Nature*, 149: 491–4.

Perutz, M. F. (1962). 'Nobel Lecture: X-ray Analysis of Haemoglobin'. *Nobelprize.org*. Accessed June 26. http://www.nobelprize.org/nobel_prizes/chemistry/laureates/1962/perutz-lecture.html.

Pietzsch, J. (2016). 'The Nobel Prize in Chemistry 1962 – Perspectives'. *Nobelprize.org*. Accessed June 26. http://www.nobelprize.org/nobel_prizes/chemistry/laureates/1962/perspectives.html.

Rheinberger, H-J. (1997). *Toward a History of Epistemic Things: Synthesizing Proteins in the Test Tube*. Stanford: Stanford University Press.

Ryle, G. (1933). 'Imaginary Objects'. *Proceedings of the Aristotelian Society Suppl.*, 2: 18–43.

Searle, J. R. (1975). 'The Logical Status of Fictional Discourse.' *New Literary History: On Narrative and Narratives*, 6 (2): 319–32.

Searle, J. R. (2010). *Making the Social World. The Structure of Human Civilization*. Oxford: Oxford University Press.

Serdyuk, I. N., N. R. Zaccai, and J. Zaccai. (2007). *Methods in Molecular Biophysics: Structure, Dynamics, Function*. Cambridge, MA: Cambridge University Press.

Smyth, M. S. and J. H. J. Martin. (2000). 'X Ray Crystallography'. *Molecular Pathology*, 53 (1): 8–14.

Snogerup-Linse, S. (2012). 'Studies of G-Protein Coupled Receptors. The Nobel Prize in Chemistry 2012. Award Ceremony Speech'. *The Royal Swedish Academy of Sciences*. Accessed 25 September 2014. http://www.nobelprize.org/nobel_prizes/chemistry/laureates/2012/advanced-chemistryprize2012.pdf.

Solid State. (2008). 'Wurtzite polyhedra.png'. *Wikipedia*. Accessed 26 June. https://en.wikipedia.org/wiki/File:Wurtzite_polyhedra.png.

Splettstoesser, T. (2006). 'X ray diffraction.png'. *Wikipedia*. Accessed 23 June. https://commons.wikimedia.org/wiki/File:X_ray_diffraction.png.

Suárez, M. (2004). 'An Inferential Conception of Scientific Representation'. *Philosophy of Science*, 71: 767–79.

Suárez, M. (ed.) (2009). *Fiction in Science: Philosophical Essays on Modelling and Idealization*. New York: Routledge.

Toon, A. (2012). *Models as Make-Believe: Imagination, Fiction and Scientific Representation*. London: Palgrave Macmillan.

Walton, K. (1990). *Mimesis as Make-Believe: On the Foundations of the Representational Arts*. Cambridge, MA: Harvard University Press.

Wimsatt, W. C. (2007). *Re-Engineering Philosophy for Limited Beings: Piecewise Approximations to Reality*. Cambridge, MA: Harvard University Press.

Winsberg, E. (2009). 'A Function for Fictions: Expanding the Scope of Science', in M. Suárez (ed.) *Fiction in Science. Philosophical Essays on Modelling and Idealisation*. London: Routledge.

5 Interpreting the sciences, interpreting the arts

Otávio Bueno

1 Introduction

Why are interpretations of scientific theories important? Why are interpretations of artworks significant? What *are* such interpretations? Which implications do they have for the ontology of fiction and the ontology of the sciences? In this paper, I begin to address these questions by contrasting different accounts of interpretation in the sciences and in the arts, and by examining the respective roles of interpretation. In both cases, I argue, interpretations provide ways of making sense of the relevant phenomena; in both cases, interpretations are connected to understanding.

Despite significant differences between the arts and the sciences, interpretations in both domains also have something in common: in some cases, they can be used to resist certain commitments, namely, in order to interpret certain things, we need to *imagine* the relevant objects rather than posit them as existing. I illustrate the resulting view by applying it to some cases from the arts (considering films in particular) and the sciences (examining some instances in chemistry and molecular biology). As will become clear, interpretations in the sciences and the arts involve a variety of considerations ranging from various ontological views to the roles played by visualization and imagination. I argue that, in many instances involving visual evidence in the sciences, the particular form of imagination that is required is precisely the one that is in place in order to make sense of key aspects of one's experience in the arts.

2 Interpretation: the sciences and the arts

Interpretations are a crucial feature of scientific practice, given that a number of significant issues that are needed to make sense of the world, and that are not directly settled by available theories, experiments, and data, are then addressed by particular interpretations of the relevant items of scientific research. Interpretations are invoked to make sense of what goes on beyond the constraints of the theoretical and empirical traits of scientific practice, but they still operate within the framework of the relevant theories and experiments and address questions that such experiments and theories fail to settle.

There is much about the world that scientific theories and experiments are in no position to resolve, but the issues in question can still be raised, since examining them helps to illuminate one's understanding of various aspects of reality. These issues include whether the world is determinist or indeterminist (van Fraassen, 1991); whether there are real modalities in nature, such as laws and objective chance (van Fraassen, 1989); what is the ultimate nature of the main constituents of the world: whether they are individuals or not, whether they have well-defined identity conditions, and what kind of things they are (French and Krause, 2006 and French, 2014). In order to examine these issues, various interpretations are advanced and defended.

Two main views regarding the nature of interpretation in the sciences should be highlighted: a *model-theoretic* conception and a *modal* view. (a) On the *model-theoretic* conception, an interpretation is an assignment of truth-values to the sentences of a theory (a *model*, in the logician's sense), so that all of its sentences come out true. This way of conceptualizing interpretations emerges from Alfred Tarski's work on formalized languages (Tarski, 1936), which eventually led, on the one hand, to the interpretation of these languages *via* a model-theoretic semantics and, on the other, to a model-theoretic conception of scientific reasoning more generally (see, for instance, Suppes, 1961 and 2002; van Fraassen, 1980; and da Costa and French, 2003).

The central features of this approach to interpretation are the notion of truth in a structure (relative to which interpretations are implemented) and the development of an overall extensional, set-theoretic approach to the understanding of scientific results. (Quasi-truth and partial structures provide a generalization of this approach; see da Costa and French, 2003.) On this conception, set theory provides the crucial mathematical framework in terms of which the interpretation of scientific results is articulated and different attitudes regarding these results in scientific practice can be represented.

(b) On the *modal conception*, an interpretation is an answer to a question regarding the way the world could be if a given theory were true. Bas van Fraassen puts the point as follows (1991, p. 9; see also van Fraassen, 1989): *How could the world possibly be how this theory says it is?* The crucial feature of this conception is that an interpretation is an exercise in *modal* reasoning. It fills out details left open by a given scientific theory as well as relevant experiments and data. In many instances, the application of a theory may require some interpretation, since the theory's conceptual resources need to be properly understood in order for them to be successfully applied. In the end, by offering details not settled by the theory, an interpretation provides understanding, since it indicates how possibilities left open by a theory (in light of experiments and data) can, at least in principle, be resolved. Note that, in this conception, the truth of the relevant theory still plays a role, since it is relative to the truth of the packages resulting from a theory plus each of its interpretations that the comparative adequacy of various interpretations is assessed.

The modal and the model-theoretic approaches to interpretation are of course connected, given the modal import that models have: models, at least on a common, reductive conception, are used as a way of (allegedly) reducing the modal content of theories to an extensional basis. On this conception, what is possible is what is true in some structure; what is necessary is what is true in all such structures.

But it is important to appreciate a significant limitation of this reductive project. In order to properly account for all possibilities, it is crucial that each possibility be representable in a model. But since models are typically formulated in set theory, such as Zermelo-Fraenkel set theory with the axiom of choice (ZFC), they inherit all of the representational limitations of this mathematical setting. For instance, it follows from this framework that it is not possible that there are objects that lack well-defined identity conditions, given ZFC's extensionality axiom. According to this axiom, sets x and y are the same as long as they have the same members, that is, provided that every member of x is also a member of y, and vice versa. In symbols:

$\forall x \, \forall y \, (x = y \longleftrightarrow \forall z \, (z \in x \longleftrightarrow z \in y))$.

Suppose that an object o lacks well-defined identity conditions. Even if o were a member of set s_1, in order to satisfy the extensionality axiom, the *same* object o needs also to be a member of set s_2, and vice versa, so that the identity of s_1 and s_2 (or lack thereof) can eventually be specified. Clearly, if some *other* object, *different* from o, albeit perhaps indistinguishable from it, were a member of s_2, this would be irrelevant for the satisfaction (or not) of the extensionality axiom as far as o is concerned. Hence, it is required that o, and, in fact, all members of the sets s_1 and s_2 in question, have properly characterized conditions of identity. Otherwise, it is indeterminate whether the extensionality axiom has been satisfied or not. As a result, this axiom ends up demanding the identity of every member of every ZFC set.

There are, however, significant interpretations of non-relativist quantum mechanics, such as those defended by Erwin Schrödinger and Hermann Weyl, according to which quantum particles (such as electrons) lack well-defined identity conditions: they are just not the kind of thing to which identity applies (for references and a thorough analysis, see French and Krause, 2006). However, on the usual model-theoretic conception, at least if it is formulated in ZFC, there is no room for such objects (those that lack identity conditions), given that the extensionality axiom always holds. Such objects are impossible in light of this conception, due to the conflict with the axiom of extensionality. But presumably one should not rule out an interpretation of quantum mechanics just because the particular set theory one adopts cannot accommodate it.

One could, of course, adopt a different framework, and set theories that do not require the identity of all the members of the collections that are formed have also been constructed, such as quasi-set theory (see French and Krause, 2006). This ends up providing additional support for the conclusion above: what is taken to be possible (or not) depends on the underlying features of the model-theoretic framework. Objects that lack well-defined identity conditions are impossible in ZFC, but they are possible in quasi-set theory. So, depending on the representational features of the underlying model-theoretic framework, different conclusions are reached regarding the relevant possibilities or impossibilities.

This does not seem to be right, though. Presumably possibilities (and impossibilities) are what they are independently of what we take them to be. The *representation* of possibilities is, of course, a different matter, and clearly it is something that is sensitive to particular frameworks. But since we are concerned with what is possible (or impossible) when we provide an account of modality, the inherent representational limitations of the model-theoretic approach raise a concern regarding its overall adequacy.

In the arts, there are also (at least) two main views of interpretation: a *metaphysical* conception, and a *making-sense* view. (i) On the *metaphysical* conception, an interpretation is whatever is ultimately required to make the claims (implicitly or explicitly) made by an artwork come out true. Many features of an artwork are often left unsettled by the work, and in order to fill out the details, an interpretation is needed. Without such an interpretation, it is indeterminate whether certain features of the work hold or not. It is indeterminate exactly how many trees surrounded Macondo in Gabriel García Márquez's *One Hundred Years of Solitude*. Nothing in the novel itself settles such an issue, nor does an answer follow from what is stated in the work. The issue is simply undecided. Now, assuming that Macondo is a relatively standard village (within the setting of the novel, in which not everything is exactly what it seems to be), it is reasonable to suppose that there is a certain number of trees surrounding the village, even if the novel is silent about exactly what that number is. Thus, to say that that number is x, for some natural number x, is to add an interpretation to the novel (admittedly, on an issue that is not pressing for the novel as a whole), given that the work itself fails to determine the truth-value of the sentence in question. By supplying this additional information, an interpretation provides details left open by the novel.

The crucial feature of this approach is that artworks have truth assignments only given an interpretation. The interpretation, by offering missing details not explicitly determined in the work, allows for truth to be assigned to what would otherwise be an incomplete, unspecified body of claims. Underlying this account of interpretation is the idea that artworks, whether paintings, photographs, dance, films, novels, short stories, or poems, make claims. Not only is there something they are like, and something they are about, but there is also something they state (or imply, suggest, or, in some cases, fail to claim) about their subject matter, even if it is contentious what that subject matter exactly is. To determine what they state typically requires an interpretation, which both fills out details not explicitly settled by the work and provides the conditions under which what is stated by the artwork is true.

But how does a painting, a photograph, or a film make claims? As opposed to novels, short stories, and poems, these art forms do not represent linguistically. Any claims they are taken to make need to be reconstructed on the basis of the perceptual content they provide. Consider, for instance, Rembrandt's self-portrait painted circa 1628 when he was 22 (it is now in the Rijksmuseum). The portrait has Rembrandt emerging from the shadows, his face partially hidden in the dark, a man who is about to emerge from obscurity, not entirely sure of what the future holds. There are salient features in the composition of the painting that clearly highlight these traits: the position of Rembrandt's body relative to the light source, which is on his back; the careful rendering of his eyes; the contour of his lips. The painting, in a straightforward interpretation, clearly states where Rembrandt stands (in the shadow), and subtly suggests his imminent motion that is likely to take him to the light in full view. The very minimal interpretation highlights what is left implicit, but can still be clearly recognized on the surface of the painting. The style is informal, with clearly visible brushstrokes, and a suggestive use of chiaroscuro.

Contrast this painting with another Rembrandt self-portrait, painted four years later in 1632 (and which is now at the Kelvingrove Art Gallery and Museum in Glasgow). Not only does the viewer see Rembrandt entirely in full light, but also the style of the painting is precisely the one that Rembrandt himself used as a very successful portraitist at the time. The painting is very formal and meticulously finished, but, despite being in its own way lovely, it lacks the intensity and immediacy of the earlier portrait. But it does depict Rembrandt as a successful, accomplished portraitist: he became his own subject in both style and form. Again, this is clearly identifiable on the painting, given the finished quality of the work, the formal tone of the composition, and the position of Rembrandt's body directly facing the viewer. Once again, in light of a very minimal interpretation, all of these traits are clearly put in view.

Finally, contrast this work with another Rembrandt self-portrait, which was painted several years later in 1659: the "Self-Portrait with Beret and Turned-Up Collar" (now at the National Gallery of Art in Washington, D.C.). Rembrandt is depicted in shadow, with the exception of his face, which is clearly illuminated: his tired eyes, wrinkles, his concerned gaze are all made salient, a man who emerged from the shadows, but who can be engulfed by darkness at any moment. The style, similar to the 1628 portrait, is less polished than the 1632 painting, with salient, vivid brushstrokes, and as opposed to the prior work, it provides a glimpse into an intimate moment of the sitter. All of this is clearly depicted on the surface of the canvas, and made explicit by a fairly straightforward interpretation of the work. In this way, claims are clearly made by all of these paintings in the context of particular (fairly minimal) interpretations of them. (The point extends in a similar manner to photographs and films, which are also perceptual-based representations and open to some minimal interpretation.)

Also underlying this conception is the proposal that fictional works are inherently incomplete (Bueno and Zalta, 2005): there are always unspecified, indeterminate features left open by the work. Interpretations fill out some of these features, indicating how certain unresolved possibilities can be completed in ways that are consistent with the extant traits of the artwork under consideration.

This is a metaphysical conception of interpretation, given that the unresolved features of the work that are examined by the various interpretations end up being resolved via the postulation of additional traits: the additions made by each interpretation, and that are not explicitly posited by the work itself. A fuller, more complete picture of the artwork in question emerges as a result.

Although this way of approaching interpretations in the arts may have some attractive features (in particular, interpretations do provide additional, and in some instances illuminating, traits to the works in question), the approach also faces some difficulties. I will stress one. Some artworks involve irony, which is not properly assessed in terms of the assignment of truth to the work in question. In fact, in the face of irony, what is explicitly stated is false and any literal interpretation of the content would clearly miss the mark.

Consider Mark Tansey's painting *The Innocent Eye Test* (Metropolitan Museum of Art in New York City). In the painting, a cow faces a painting of cows, which has just been revealed to it. The draperies that had been covering the painting rest loosely on the ground as scientists and critics carefully observe and take note of the cow's reaction. Does the cow recognize the painting in front of it as of a cow (or, less likely, as the painting of a cow)? Can the critics and scientists studying the cow fully apprehend what goes on in the cow's experience? The setting of the scene is that of a museum, with another painting, Monet's *Grainstack* (*Snow Effect*), positioned at the side of the painting of the cows and visible on the wall. Tansey's painting raises a number of issues, including: How can one determine the extent to which the world is represented differently by different creatures? What kind of recognition should be given to perfectly realistic paintings? How artificial is the setting of a museum as a place to experience art?

That the painting raises all of these issues (among others) is in no doubt. However, the painting can also be interpreted as providing an ironic commentary on the state of the arts, in which critics, with the presumed objectivity of scientists, aim to find out the true nature of artistic representation, but are just as bewildered as scientists who try to figure out the mental states of cows. Moreover, many viewers relate to the artworks they experience just as the cow in Tansey's painting presumably relates to the painting of cows: both, viewers and cow, blankly stare at the paintings in front of them, chewing, oblivious to their surroundings and the paintings they stare at.

It is unclear, however, how such an irony can be captured in terms of assignments of truth-values to relevant artworks. The viewers of Tansey's painting certainly are not depicted in the painting, and thus additional objects have to be posited by the interpretation to accommodate the work. This is expected on this account of interpretation. But how can the irony of the situation of the viewers relative to the cow be captured? It may be argued that the irony is accommodated by identifying the structural similarity between, on the one hand, the situation of the cow and the painting of cows, carefully observed by critics and scientists in Tansey's painting, and, on the other hand, the situation, in an actual museum, of the viewer and the painting that is being viewed (namely, Tansey's painting of a cow observing the painting of cows). These two situations are indeed structurally similar: they both involve the relation of being in front of a painting and observing it. But the structural similarity between a cow that observes a painting of cows and a viewer that observes a painting of a cow (observing a painting of cows), despite the structural match between them, does not capture the *irony* of the painting. The structural similarity between the relevant situations, interpreted as something that is true, fails to make the crucial point, which is the commentary about the viewers as cows, and the critics as hopelessly lost. Viewers are considered to be creatures just as oblivious to paintings as cows, and critics (as well as scientists) who study the painting (and the cow) have as much grasp of what is going on regarding the actual situation of the arts as the scientists and critics do in Tansey's own painting.

None of this is a matter of identifying true claims in the artwork. Quite the contrary, the irony of the painting requires that viewers be thought of as cows without ever being themselves cows, without ever being mapped into cows. A literal, structural account of the situation would undermine the irony of the work. What we have here is, rather, a matter of making sense of the artwork in ways that are compatible with the content that is actually displayed in it, while allowing for what is stated in the work to be interpreted as, in fact, stating something else altogether.

(I do not deny that structural preservation plays a central role in making sense of artworks. On the contrary, such structural relations are really important (see Bueno and French, 2011). What I am questioning is the claim that truth, even in the context of an interpretation, is the proper requirement for the assessment of artworks. Understanding is a much more significant constraint and goal.)

These considerations motivate a second way of approaching interpretation in the arts. (ii) On the *making-sense* conception, an interpretation provides an account that ultimately makes sense of the artworks in question. These works, in addition to leaving a number of features incomplete, often display or suggest (intentionally or not) a number of puzzling traits and unresolved tensions. To address these issues, interpretations are articulated, and similarly to the metaphysical conception, they do add extra traits to the original work, and in ways that are consistent with its content. However, as opposed to the metaphysical conception, truth is not a requirement for an interpretation. On the making-sense conception, the central feature of an interpretation is to resolve, or at least highlight, the tensions and puzzling features of artworks, particularly those that raise significant issues regarding the meaning and the content of the work. By engaging directly with issues that hinge on the proper account of an artwork, and by emphasizing the importance of making sense of the work as part of the point of putting forward an interpretation, the making-sense conception provides understanding – quite independently of any truth assignment to the artwork in question. Rather, the overall approach turns on probing the features of the work under consideration, adding possible details as needed to make sense of and resolve (whenever possible) the tensions and complexities that are embedded in the artwork.

The making-sense conception of interpretation offers an account that is less artificially constrained than the metaphysical conception. Interpretations are advanced *via* any resources that are needed or may be available, provided that they provide understanding. One is not tied to the requirement that the claims involved in the works in question given an interpretation need to be true. However, one is constrained by the content of the work, and what is added by an interpretation should be compatible with that content. Otherwise, instead of an interpretation of the target artwork, one would end up considering a work that is fundamentally different from the original, namely, a work in which the negation of certain claims in the original artwork are made. After all, it is that revised work that one will be interpreting if what is added by an interpretation is incompatible with the original work: an interpretation should minimally be consistent with the work being interpreted. (I take it that the package of an artwork plus an interpretation should be coherent, even if the original artwork is not consistent. Coherence does not require consistency (see Bueno and da Costa, 2007): Frege's original logicist reconstruction of arithmetic, despite being inconsistent, was certainly coherent.)

This allows for irony, sarcasm, parody, satire, pastiche, and a variety of modes of expression that are not literal renditions or faithful representations of the actual world to be accommodated. The goal is to understand – to make sense of – what an artwork conveys, its central points, and how the specific resources available in the medium of the work are employed in order to fully express these points, while resolving, along the way, tensions and puzzles that are raised by the work. The central points expressed by an artwork are closely tied to the story an artwork conveys.

Every story always involves two stories (Piglia, 2011): one explicitly told and another hidden, never overtly articulated. It is out of the interplay between the two stories that the full story ultimately emerges. As Roberto Piglia notes:

> In one of his notebooks, Chekhov recorded the following anecdote: 'A man in Monte Carlo goes to the casino, wins a million, returns home, commits suicide.' The classic form of the short story is condensed within the nucleus of that future, unwritten story. Contrary to the predictable and conventional (gamble–lose–commit suicide), the intrigue is presented as a paradox. The anecdote disconnects the story of the gambling and the story of the suicide. That rupture is the key to defining the double character of the story's form. First thesis: a short story always tells two stories.
>
> (Piglia, 2011, p. 63)

The double-story model allows for the creative exploration of different connections between the explicit and the hidden stories ('Story One' and 'Story Two', respectively, in Piglia's terminology). The hidden story, concealed throughout as the explicit story is told, can only surface at the very end, when it is realized that it has been underlying the explicit story from the start. These, on Piglia's account, are the classic short stories.

> The classic short story—Poe, Quiroga—narrates Story One (the tale of the gambling) in the foreground, and constructs Story Two (the tale of the suicide) in secret. The art of the short story writer consists in knowing how to encode Story Two in the interstices of Story One. A visible story hides a secret tale, narrated in an elliptical and fragmentary manner. The effect of surprise is produced when the end of the secret story appears on the surface.
>
> (Piglia, 2011, p. 63)

The tensions between the explicit and the hidden stories are, in some cases, resolved, but in others, such as in the modern version of short stories, a resolution is never achieved. The story articulates the antagonism between the two stories without ever reaching a solution.

The two models (classic and modern) can also be combined:

> The classic short story *à la* Poe told a story while announcing that there was another; the modern short story tells two stories as if they were one. Hemingway's 'iceberg theory' is the first synthesis of that process of transformation: the most important thing is never recounted. The secret story is constructed out of what is not said, out of implication and allusion.
>
> (Piglia, 2011, pp. 64–65)

The allusion and implication allow the hidden story to emerge, albeit implicitly, since it is never fully stated. Moreover, no resolution between it and the explicit story is ever articulated, given that the implicit story never becomes overt.

In light of the dynamic character of the relation between the hidden and the explicit stories, interpretation plays a key role: it is a matter of interpretation to make sense of what the proper relation between the two stories ultimately is. It is thus tempting to consider the hidden story as offering the underlying point of the overall story. Curiously, Piglia insists that the hidden story is not a matter of identifying an underlying point that relies on interpretation. Rather, he tells us, the hidden story is simply told in a mysterious way.

> The short story is a tale that encloses a secret tale. This is not a matter of a hidden meaning which depends on interpretation: the enigma is nothing other than a story which is told in an enigmatic way. The strategy of the tale is placed at the service of that coded narration. How to tell a story while another is being told? This question synthesizes the technical problems of the short story. Second thesis: the secret story is the key to the form of the short story.
>
> (Piglia 2011, p. 64)

On Piglia's view, the dynamic between the hidden and the explicit stories is at the core of storytelling in the case of short stories, and that is the primary reason to focus directly on the two kinds (explicit and hidden) of stories rather than on the role of interpretation.

However, despite Piglia's own theoretical preference to highlight the role of the double stories, the model does require that a key role be assigned to interpretation. In order to identify exactly how the hidden stories emerge through the interstices of the explicit story, one needs to figure out the point of the story, and that, of course, requires interpretation. Moreover, despite Piglia's own remarks, the hidden story can indeed be thought of as the key point of the narrative, that which is suggested, implied, insinuated, or hinted at, but never fully stated. More than a narrative device, the very structure of a story requires the identification of the hidden story, but this, in turn, demands an interpretation: it demands that one make sense of the interplay of the overt and the hidden stories. So Piglia's own account, in the end, relies on assigning a crucial role to interpretation. And the making-sense conception accommodates this role very well.

But as opposed to the metaphysical conception, the making-sense conception does not provide a formal framework in terms of which interpretations are advanced, and it may be considered, for this reason, to be too open-ended. A formal framework often helps to provide structure and conceptual resources to accommodate various issues that need to be addressed when interpretations in the arts are in question. In response, the concern can also be raised that some formal frameworks introduce artificial distinctions that have very little to do with the content of what needs to be interpreted and are not very well-connected with the way the relevant practices are articulated. So there are worries on both sides.

However these issues are resolved, we have here two distinct approaches to interpretation in the arts. Each of them has their virtues, and each faces some difficulties. A form of pluralism of interpretations, which acknowledges the benefits offered by each account while also recognizing their limitations, is recommended.

3 Ontology: the sciences and the arts

Matters of interpretation in both the sciences and the arts often involve ontological issues. As noted above, in some cases, an interpretation may involve the postulation of additional objects and relations beyond those explicitly posited by a given theory or practice. How should this additional ontology be accommodated? Once again, I will start with the case of the sciences, and then move to the arts.

Two broad approaches to the ontology of scientific theories have been formulated: ontologically committing conceptions and ontologically noncommitting ones. (a) According to *ontologically committing views*, we ought to be ontologically committed to all and only those objects (including relevant properties, relations, and structures) that are indispensable – in predictive, explanatory and expressive contexts – to our best theories of the world. These are the objects that are taken to exist. They include concrete objects, such as electrons, positrons, and quarks, as well as abstract entities, such as classes (Quine, 1960; Putnam, 1979, 2012a; and Colyvan, 2001). It is the indispensability of the objects that are quantified over, in order to account for, explain or express a variety of phenomena (those aspects of the world that need to be accommodated), that ultimately leads to the ontological commitment to the objects in question. The indispensability of such objects in the various contexts in which they are used entails that theories that quantify over them cannot be reformulated without referring to these objects on pain of involving significant loss in predictive, expressive, or explanatory power. Quantification over the relevant objects in service of scientific achievements, thus, becomes indispensable, and is taken to be a mark of the real.

(b) In contrast, according to *ontologically non-committing views*, quantification is not enough to guarantee ontological commitment, even when it is indispensable to our best theories of the world. If quantifiers are interpreted as being ontologically neutral (Azzouni, 2004; Bueno, 2005), the sheer fact that a scientific theory involves quantification over a given domain of objects is not enough to guarantee ontological commitment – that is, commitment to the existence – of these objects, even if reference to or quantification over them turns out to be indispensable to the theories in question.

In order to express ontological commitment, an existence predicate is added to the language. Different sufficient conditions for the satisfaction of the existence predicate ("conditions of existence" for short) can then be specified, depending on the particular ontological views one favors. One may defend the account according to which ontological independence (from linguistic practice and psychological processes) provides such conditions of existence (Azzouni, 2004). In this view, what exist are those things that do not depend ontologically on the particular vicissitudes of our speeches and thoughts. Jody Azzouni takes the ontological independence condition also to be necessary, but I do not see how this move can be implemented without begging the question against platonism, which is something he aims to avoid (see Bueno, 2013).

Alternatively, one may argue that certain forms of access to objects provide such sufficient conditions of existence. This is the case, for instance, of an access that is sensitive to features of the objects in question such that (i) had the objects been different (within the sensitivity range of the relevant access), the resulting experiences would have been correspondingly different, and (ii) had the objects been the same (again, within the sensitivity range of the relevant access), the resulting experiences would have been correspondingly the same (see Bueno, 2011). This form of access, which includes both perception and instrumentally aided observation, thus establishes a counterfactual dependence between the objects in question and the relevant experiences, whether they are perceptual experiences or instrumentally mediated ones. The access yields a control over objects that has a number of epistemically relevant features: it is robust, stable, and can be refined (for additional discussion of these traits, see Azzouni, 2004).

In the arts, there are also two main views regarding the status of fictional entities, such as Hamlet or Sherlock Holmes. (a) According to the ontologically committing view (a particular form of realism about fictional entities), fictional objects do exist. These objects have perfectly well-defined existence and persistence conditions, particularly in the context of literature (Thomasson, 1999). There is no mystery as to how fictional entities come into being (how they come to exist) and how they continue to exist (their persistence conditions). All it takes for a fictional entity to exist is for appropriate intentional acts of an author to be invoked: Conan Doyle's descriptions of the deeds of a detective who lived in London and solved crimes brilliantly are enough for the corresponding object to come into existence. Nothing else is really needed.

None of the descriptions in question are complete, nor can they be: there are always features of fictional objects that are left unspecified. Nothing in the story establishes, for some of these features, what they are, nor does it follow from what is explicitly stated in the story that the relevant objects have (or lack) certain properties. In fact, as I noted above, being incomplete seems to be a distinctive, characterizing feature of fictional objects.

As for the persistence conditions, they are satisfied as long as stories figuring the relevant fictional characters are told, remembered, or recalled. If no copies of the works in question are preserved, if no memories of the characters remain, the fictional objects under consideration will no longer exist (Thomasson, 1999).

On this view, fictional characters are then *abstract artifacts*. They are *artifacts* since they are made up by the intentional acts of authors, and they are *abstract* since they are non-spatial objects. They are, however, temporal entities, coming into existence at certain moments in time and getting out of existence at others.

(b) According to the ontologically non-committing view, fictional objects do not exist or, at least, need not be taken to exist for us to make sense of our literary and artistic practices. That is, non-ontologically committing views regarding fictional entities can be either strictly nominalist (thus involving the denial of the existence of the relevant entities) or agnostic (thus avoiding the commitment to the entities in question, while remaining neutral regarding their existence or not).

One straightforward way of implementing either strategy is by invoking ontologically neutral quantifiers (Azzouni, 2004; Bueno, 2005). As noted above, with these quantifiers in place, quantification even over objects that are indispensable to the prediction, expression, or explanation of certain phenomena is not ontologically committing. If a necessary *and* sufficient condition for the existence of something is that it is ontologically independent from our linguistic practices and psychological processes (Azzouni, 2004), a nominalist view about fictional entities is obtained. After all, the argument goes, fictional objects are entirely made up by us, and thus their existence depends on the intentional acts of those who have created them. Since the ontological independence condition is violated, it follows that fictional objects do not exist. (Interestingly, as we saw, ontologically committing views would draw precisely the opposite conclusion from the same assumptions. That is why an agnostic view should be preferred on these issues: it is unclear how they could be resolved in a principled way in any case.)

Alternatively, if a *sufficient* (but not necessary) condition for the existence of something is that it is counterfactually dependent on certain forms of access to it, an agnostic view about fictional objects emerges. Since the condition is only sufficient, nothing can be said about the nonexistence of fictional entities should the condition *not* be satisfied. Moreover, given that arguably fictional entities do not satisfy the condition (they are not counterfactually dependent on certain forms of access), nothing could be said about their existence either. The resulting view is then a form of agnosticism about fictional entities. Since neither the nominalist nor the agnostic views are committed to the existence of these entities, there is no need to provide existence or persistence conditions for them.

Having presented these different approaches to ontology, we can now examine their implications to the ontology of fiction. As an illustration, I will consider, in particular, films, given the interesting ways in which films bring together pictorial (perceptual) and linguistic (descriptive) traits. As will become clear, a crucial role is played by interpretation and imagination in this context.

4 Seeing and imagining in fiction and film

In the case of films, one issue to be addressed is how to make sense of our experience of watching them, and the fact that we seem to be seeing fictions in such films. There are views regarding the ontology of film that correspond to the ontological views just discussed. (a) According to the ontologically committing view, all it takes for one to see a fictional character in a film is for one to watch the screening of the film in which the relevant character is shown. Thus, existence conditions for fictional objects in film are straightforward. Similarly, as long as the films continue to be screened, the corresponding persistence conditions are also satisfied. On this view, fictional objects in film are abstract artifacts. But how can one *see* an *abstract* object? How can one see an object that is not in space?

(b) According to the ontologically non-committing view, since fictional characters do not exist (or, at least, we need not to take them to exist), we cannot see them. Nonexistent objects cannot be seen. The issue then becomes one of making sense of the phenomenology of watching a film. What is seen on the screen if not the characters from the narrative that is presented in it (when that is what the film portrays)? It seems natural to describe our experience of watching films as one in which we see the characters on the screen. The difficulty, however, is then of explaining how exactly we can see a nonexistent object.

Making sense of our phenomenology of watching films is a crucial feature of a philosophical account of them, whether one adopts an ontologically committing view or a non-committing one. It is in this context that interpreting films become particularly significant. It involves making sense of the visual content that is portrayed on the screen, as part of the larger project of making sense of the film and what it conveys. As will become clear, a particular kind of imagination – an imaginative act of perceptual imagination – is crucial here.

Films are watched by being interpreted, by making sense of what is going on in the images displayed on the screen. What are these images of? Excluding computer-generated films, these are images of the actors and the sets that were in front of the camera when the film was being shot. When Harrison Ford was working in the shooting of *Blade Runner*, the images of his performance resulted from his interaction with the camera. His presence in front of the camera caused the images that eventually ended up in the film. The images, in turn, given the way in which they were created, represent him. They are images of Harrison Ford. But clearly viewers of the film had better not watch *Blade Runner* thinking that these are images of the actor, although literally that is what they are. Otherwise, it will be very difficult to get access to the film's content, which is telling a story not about Harrison Ford but about a fictional character, Deckard, in his search for replicants (and whether or not he is one of them). But the issue returns: how can one see a fictional character in a film?

As we saw, on the ontologically committing view, fictional characters do exist, and they are abstract artifacts. As noted above, their abstract character emerges from the fact that, despite being temporal objects, abstract artifacts are not spatially located. This makes it hard to account for how we can see them. But here is one way in which this issue can be addressed. Perhaps all it takes for a fictional object to exist in a film is for the film to be screened. The character is constituted by the sequence of shots and the narrative structure that comprises the film. In order for one to see a fictional character in a film, it is enough to see the corresponding images on the screen (perhaps this is the view advanced in Wilson, 2011).

However, what one literally sees on the screen are images of the actors, and the characters are not images. In *Blade Runner*, Deckard is not an image of Harrison Ford, but rather a person (or, perhaps, a replicant). Given the transparency of photographic and cinematic images (the counterfactual dependence between the recorded image and the object that is being photographed or shot), one can understand how by seeing an image of Harrison Ford, one can see Ford. But what is required by the ontologically committing view is that *Deckard*, and not Ford, be seen. How is this achieved? Perhaps it could be argued that all it takes to see Deckard is to see Ford acting on the screen. Ford's image, in the context of the film, just is Deckard. But this is not quite right, given that Deckard, as noted, is not an image, but a man (or, perhaps, a replicant). It then becomes unclear how, on the ontologically committing view, one can literally see characters on films.

On the ontologically non-committing view, the transparency of films – between the objects and their corresponding images – also holds. On this view, images are generated from the interaction of the camera with the objects that are filmed, and can be taken to represent the objects these images are about. But according to this proposal, fictional characters do not exist (or need not be taken to exist). So how can these nonexistent objects be seen?

They cannot, and are not. Fictional characters are not seen, but they are *imagined* based on the perceptual content given by the film, which is, primarily, conveyed by the images on the screen. (I believe this is a view favored by Currie, 1995.) Although we cannot see something that does not exist, we have no trouble imagining nonexistent objects. What is involved in film is a special type of imagining, an imaginative act guided by the perceptual content displayed on the film: *perceptual imagining*. (Scruton, 1979: Chapter 4, provides an illuminating discussion of the role of perceptual imagining in architecture.) When we watch *Blade Runner*, we see an image of Harrison Ford, and imagine Deckard, not as a fictional character, but as a person (or, perhaps, as a replicant). The content of the character (his facial expressions, demeanor, and attitude, as well as his looks), as imagined by us based on the images on the film, are constrained by the visual experiences of seeing what is on the screen. The eye color, mannerisms, and gestures of the actor inform and constrain our imaginative act.

Perceptual imagining provides a particular form of interpretation, a particular way of seeing the actor *as* the character the image on the film represents. It allows viewers to make sense of the film, experiencing the image on the screen *as* an embodiment of the character it stands for. Since the nonexistence of characters (on the non-committing view) provides no hindrance to the imagination, and given the perceptual content involved in the experience of a film, with perceptual imagination we have the combination of interpretation with perceptual content.

Clearly, the imaginative acts involved in perceptually imagining Deckard by watching *Blade Runner* and by simply imagining him by reading the accounts provided in Philip K. Dick's novel *Do Androids Dream of Electric Sheep?* are very different. In the latter, there is no perceptual content to constrain the experience; imagination is guided by the linguistic descriptions plus any additional content provided by the reader. In the former, the imagination is entirely constrained by the perceptual content provided by the film, with correspondingly less content provided by the viewer. (Despite the difference, there is still much left to be interpreted and to be provided by readers and viewers as they make sense of the novel and the film, particularly given that the very content of these works is interestingly open to a variety of interpretations.)

Note that the two kinds of imaginings invoked here are equally familiar. *Simple imagining* – that is, to imagine something that is not perceptually present – corresponds to the usual kind of imagination. But *perceptual imagining* – that is, to imagine of something that is perceptually present that it is something else – is no less common, and is also perfectly legitimate. Suppose I am trying to show you, using my fork, knife, and spoon, the relative positions of Miami, Orlando and NYC. I say, grabbing the fork and placing it on the table, "Imagine this is Miami." I then grab the knife and place it above the fork. "Orlando would be here," I say, and, placing the spoon well above the knife, add, "NYC would be way up there." To imagine the relative positions of Miami, Orlando, and NYC based on the visual experience of the relative positions of the fork, knife, and spoon is to engage in a process of perceptual imagining. One sees the relative positions of the cutlery and imagines the relative positions of the cities: a perfectly familiar imaginative act.

Both imaginings are involved in experiencing films. Consider the difference between, upon watching *Blade Runner*, (a) imagining that Deckard fell in love with Rachel or that he is a replicant, and (b) imagining that Deckard was able to fly. We can say that (a) is an instance of *perceptual imagining*, given the visual content of the film. His love for Rachel is clearly suggested in the film; whether he is a replicant is interestingly left more open. In contrast, there is nothing in the film that supports (b). Deckard does not have superpowers (even though he may be a replicant). One can, of course, simply imagine that he has such capacity (and never actualizes it), but it does not amount to a properly supported interpretation of the film. Interpretation, guided by perceptual imagination, is constrained by the visual content of the film.

5 Seeing and imagining in the sciences

Perceptual imagining, as involved in the interpretation of scientific results, also plays an important role in the sciences. Consider, for instance, the use of the transmission electron microscope by George Palade (1955). When, in the early 1950s, he stumbled upon a number of dots scattered throughout the membranes of the cells that he was studying, and which were clearly visible on the images the electron microscope generated, he had no clear idea of what exactly he was seeing. He initially thought that the dots could be artifacts of the method of preparation of the sample. But, after changing the methods, the dots were still present in the resulting images. He then varied the animals from which the cells were being collected, and the parts of the body the cells came from. The dots were still there. At this point, he suspected that the dots were not an artifact of the experiment, but perhaps a genuine new cellular structure. After ultra-centrifuging the sample, and noting that the observed decantation time of the material was compatible with the measured size of the dots (which was done by using the data provided by the electron microscope), he concluded that a new cellular structure had been found.

Looking at the images of the cell and the multitude of dots near the membranes depicted on the surface of the image generated by the electron microscope, Palade did not know what exactly he was seeing. (His situation was not very different from the one the early users of optical microscopes faced when they saw microorganisms without knowing exactly what they were seeing.) He imagined that he was seeing an acid, but was unsure whether this was indeed the case. He had no idea of the function or the exact structure of those dots, and when he finally published the paper describing the results of his experiments and displaying the images he obtained, he gave it the least committing title possible: "A Small Particulate Component of the Cytoplasm" (Palade, 1955). A few years later, additional chemical analysis was implemented, and it was determined that these particulate components were indeed primarily composed of ribonucleic acid. They were then called "ribosomes."

In order to imagine certain features of the ribosomes, Palade was being guided by the perceptual information provided by the electron microscope. In this process, he was invoking a form of perceptual imagination that is similar to the one involved in the experience of films. In both cases, the perceptual content provided by the images constrains and shapes the imagination of the relevant objects. The objects are imagined on the basis of how they are perceptually experienced, and the resulting interpretations are formulated taking into account the perceptual imagination in question.

As an additional example, consider the study of the surface of structures at the nanoscale, using scanning tunneling microscopes (STMs) or atomic force microscopes (AFMs; Chen, 1993; Bueno, 2011). The data that are generated by these instruments provide information about objects that, however, is not sufficient to perceptually characterize them. After all, the same data collected by the instrument regarding the samples are compatible with perceptually different images. It is then unclear whether the counterfactual conditions between the samples under study and the resulting images are known to be satisfied. (In this respect, these instruments, AFMs and STMs, are importantly different from electron or optical microscopes.) The objects in question are unobservable and cannot be literally seen – even *via* the relevant instruments. But STMs and AFMs allow researchers to *imagine* the unobservable, a sort of imagination constrained, in part, by the data provided by these instruments, which, thus, help researchers make sense of the objects being studied. Here, as elsewhere, imagination and interpretation go hand in hand.

As a final example, in molecular biology, despite the clear goal of understanding the behavior of live organisms, biologists are often limited, given the constraints of available instruments, to studying dead specimens. The relevant samples, which are artificially prepared so that they can be studied within the constraints of the instruments' specifications, not only are not alive, but they typically provide only a fraction, a slice, of the entire organism being investigated. It is, of course, far more manageable to examine just some isolated parts of the entire organism, leaving a fuller picture to a later moment once enough partial information is available. But significant distortions can also be introduced as a result of this strategy of dividing up the larger problem into manageable parts. A significant role is left to the imagination and interpretation given that not enough information is actually available about the samples under study. What is ultimately needed are better imaging instruments of live processes at the molecular level, which provide evidence for those traits that are only imagined (or inferred) from the available partial information that data from dead samples provide (see Boulina *et al.*, 2013 for a promising development in this direction).

6 Conclusion

Interpretations play important roles in the sciences and in the arts. In both cases, they provide ways of making sense of the relevant phenomena. In both cases, they are connected to understanding. They also suggest ways in which one can resist certain commitments, given those cases in which, in order to interpret certain phenomena, we need to *imagine* the relevant objects rather than posit them as existing. Whether one considers the sciences or the arts, interpretations, often guided by perceptual imagination, are crucial and ubiquitous.

Acknowledgments

For extremely helpful discussions on the issues examined in this paper, my thanks go to Ann-Sophie Barwich, Allan Casebir, Akira Chiba, George Darby, Victor Durà-Villà, Catherine Elgin, Steven French, John Kulvicki, Dom Lopes, Aaron Meskin, Dean Rickles, Jordan Schummer, and Nola Semczyszyn.

References

Azzouni, J. (2004). *Deflating Existential Consequence: A Case for Nominalism*. New York: Oxford University Press.

Boulina, M., H. Samarajeewa, J. Baker, M. Kim, and A. Chiba. (2013). 'Live Imaging of Multicolor-Labeled Cells in *Drosophila*'. *Development,* 140, pp. 1605–13.

Bueno, O. (2005). 'Dirac and the Dispensability of Mathematics'. *Studies in History and Philosophy of Modern Physics,* 36, pp. 465–90.

Bueno, O. (2011). 'When Physics and Biology Meet: The Nanoscale Case'. *Studies in History and Philosophy of Biological and Biomedical Sciences,* 42, pp. 180–9.

Bueno, O. (2013). 'Nominalism in the Philosophy of Mathematics'. in Edward N. Zalta (ed.), *Stanford Encyclopedia of Philosophy* (Fall 2013 edition). Available at: http://plato.stanford.edu/archives/fall2013/entries/nominalism-mathematics/.

Bueno, O. and Zalta, E. (2005). 'A Nominalist's Dilemma and Its Solution'. *Philosophia Mathematica,* 13, pp. 294–307.

Bueno, O. and da Costa, N.C.A. (2007). 'Quasi-Truth, Paraconsistency, and the Foundations of Science'. *Synthese,* 154, pp. 383–99.

Bueno, O. and French, S. (2011). 'How Theories Represent'. *British Journal for the Philosophy of Science,* 62, pp. 857–94.

Chen, J. (1993). *Introduction to Scanning Tunneling Microscopy*. New York: Oxford University Press.

Colyvan, M. (2001). *The Indispensability of Mathematics*. New York: Oxford University Press.

Currie, G. (1995). *Image and Mind: Film, Philosophy, and Cognitive Science*. Cambridge: Cambridge University Press.

da Costa, N.C.A. and French, S. (2003). *Science and Partial Truth: A Unitary Approach to Models and Scientific Reasoning*. New York: Oxford University Press.

van Fraassen, B. C. (1980). *The Scientific Image*. Oxford: Clarendon Press.

van Fraassen, B. C. (1989). *Laws and Symmetry*. Oxford: Clarendon Press.

van Fraassen, B. C. (1991). *Quantum Mechanics: An Empiricist View*. Oxford: Clarendon Press.

French, S. (2014). *The Structure of the World: Metaphysics and Representation*. Oxford: Oxford University Press.

French, S. and D. Krause. (2006). *Identity in Physics: A Historical, Philosophical, and Formal Analysis*. Oxford: Oxford University Press.

Palade, G. (1955). 'A Small Particulate Component of the Cytoplasm'. *Journal of Biophysical and Biochemical Cytology*, 1, pp. 59–79.

Piglia, R. (2011). 'Theses on the Short Story'. *New Left Review*, 70, pp. 63–6.

Putnam, H. (1979). *Mathematics, Matter and Method*. Philosophical Papers, Volume 1. (Second edition.) Cambridge: Cambridge University Press.

Putnam, H. (2012a). 'Indispensability Arguments in the Philosophy of Mathematics', in Putnam (2012*b)*, pp. 181–201.

Putnam, H. (2012b). *Philosophy in an Age of Science: Physics, Mathematics, and Skepticism*. (Edited by Mario De Caro and David Macarthur.) Harvard, MA: Harvard University Press.

Quine, W.V. (1960). *Word and Object*. Cambridge, MA: The MIT Press.

Scruton, R. (1979). *The Aesthetics of Architecture*. Princeton, NJ: Princeton University Press.

Suppes, P. (1961). 'A Comparison of the Meaning and Uses of Models in Mathematics and the Empirical Sciences', in Freudenthal (ed.) London: Springer, pp. 163–77.

Suppes, P. (2002). *Representation and Invariance of Scientific Structures*. Stanford: Center for the Study of Language and Information, CSLI Publications.

Tarski, A. (1936). 'Der Wahrheitsbegriff in den formalisierten Sprachen'. *Studia Philosophia* 1, pp. 261–405. (English translation, 'The Concept of Truth in Formalized Languages', in Tarski [1983], pp. 152–278.)

Tarski, A. (1983). *Logic, Semantics, Metamathematics*. (Translated by J.H. Woodger. Second edition edited by John Corcoran.) Indianapolis: Hackett.

Thomasson, A. (1999). *Fiction and Metaphysics*. Cambridge: Cambridge University Press.

Wilson, G. (2011). *Seeing Fictions in Film: The Epistemology of Movies*. Oxford: Oxford University Press.

6 Deep indeterminacy in physics and fiction

George Darby, Martin Pickup and Jon Robson

1 Introduction

1.1 Indeterminacy

Indeterminacy in its various forms has been the focus of a great deal of philosophical attention in recent years. Much of this discussion has focused on the status of vague predicates such as 'tall', 'bald' and 'heap'. It is determinately the case that a seven-foot person is tall and that a five-foot person is not tall. However, it seems difficult to pick out any determinate height at which someone becomes tall. How best to account for this phenomenon is, of course, a controversial matter. For example, some (such as Sorensen, 2001 and Williamson, 2002) maintain that there is a precise height at which someone becomes tall and such apparent cases of indeterminacy merely reflect our ignorance of this fact. Others maintain that there is some genuine – and not merely epistemic – indeterminacy present in such cases and offer various accounts of how best to account for it. Supervaluationists (such as Keefe, 2008), for example, claim that the indeterminacy with respect to vague terms lies in their not having a single definite extension. Rather, each term is associated with a range of possible precise extensions or precisifications, such that it is semantically unsettled which is the correct extension. One precisification of 'tall' might allow that anyone over five feet, ten inches is tall, whereas another would only allow those over six-foot to qualify; but no precisification will take someone who is five foot to be tall, and someone who is seven foot will count as tall on all precisifications. Thus – while someone who is seven foot will be determinately tall and someone who is five foot determinately not so – it will be indeterminate whether someone who stands at five foot, eleven inches is tall.

Yet, it is important to stress that putative cases of indeterminacy are not limited to vague predicates of this kind. Philosophers have invoked indeterminacy in discussions of topics as diverse as moral responsibility (Bernstein, forthcoming), identity over time (Williams, 2014) and the status of the future (Barnes and Cameron, 2009). In this paper, we focus on two areas where discussion of various kinds of indeterminacy has been commonplace: physics and fiction. We propose a new model for understanding indeterminacy across these domains and argue that it has some notable advantages when compared to earlier accounts. Treating physics and fiction cases univocally also indicates an interesting connection between indeterminacy in these two areas.

In the remainder of this section, we briefly survey some issues concerning indeterminacy in our target domains before introducing influential models for addressing them. In Section 2 we argue that, as things stand, these models are unable to account for a certain kind of indeterminacy which we label 'deep indeterminacy'. Section 3 introduces our new model for indeterminacy before showing how it can be applied across our two domains. In particular, we highlight ways in which it is able to account for the kinds of deep indeterminacy we introduced in Section 2. In Section 4 we offer some concluding remarks.

1.1.1 Metaphysical indeterminacy

Metaphysical indeterminacy is indeterminacy that is non-representational. This contrasts it with the vague predicate cases discussed above. Metaphysical indeterminacy is vagueness in the world, not just vagueness in our representation of the world. But the intuitive idea of metaphysical indeterminacy as indeterminacy that is non-representational requires elaboration. One line here is that metaphysical indeterminacy is that indeterminacy that would remain even if all of the representational indeterminacy were removed – that is, if the language were completely precisified. Questions about what metaphysical indeterminacy is supposed to be intersect, of course, with the question of whether there really is, or could be, any. A sceptical tradition famously writes off the idea as incoherent, or only coherent enough to be seen to be impossible or similar.

In response to that scepticism, metaphysicians have recently been constructing explicit models of metaphysical indeterminacy. Examples include Akiba (2004), Barnes and Williams (2011) and Wilson (2013). Here we focus on a particular class of models along the lines of those that Wilson calls 'meta-level' accounts (in contrast to her own 'object-level' view). The models we are interested in have a form analogous to linguistic indeterminacy as alluded to above: Where language could be more precise, but is not, with something's being indeterminate if true on some precisifications of the language and false on others, so the metaphysical version has it that the world could be more precise, but is not, with something's being indeterminate if true on some precisifications of the world and not on others. We will focus on one particular meta-level model of metaphysical indeterminacy: the one proposed by Barnes and Williams (2011). Barnes and Williams (BW hereafter) use possible worlds to give their account: metaphysical indeterminacy is to be understood by using sets of possible worlds that differ with respect to the truth of some propositions. The members of these sets of possible worlds are those worlds that are candidates for being the way the world really is. The propositions they do not agree on are the propositions that are metaphysically indeterminate. The BW model therefore treats metaphysical indeterminacy in much the way that supervaluationists treat representational indeterminacy. The details of the BW approach will be discussed in Section 1.1.3. First we introduce fictional indeterminacy.

1.1.2 Fictional indeterminacy

Before we can discuss fictional indeterminacy, we need to briefly consider some more general issues concerning the nature of fictions. In particular, we need to consider the question of what is required for some claim to be true within a certain fiction. It is a commonplace that not everything that is true in a particular fiction is explicitly stated in the fiction itself. Consider, for example, that the majority of fictional characters we encounter are more or less ordinary human beings. They lack the ability to fly under their own power, they have never engaged in exotic practices such as time travel or astral projection and they breathe oxygen rather than methane or gold dust. Yet it is exceedingly rare for an author to take the time to explicitly highlight any of these features (rather, it is the exceptions that tend to warrant particular notice). This much is generally taken to be uncontroversial (though D'Alessandro (forthcoming) questions this orthodoxy), but disputes quickly arise as to precisely what account we should give of something's being true in a given fiction. According to Greg Currie (1990: 80) what is true in a fiction is what 'it is reasonable for the informed reader to infer that the fictional author' of the fiction in question believes.[1] Derek Matravers (1997: 423) takes it to be a matter of 'what it would be reasonable for the reader to infer (within the game of imagining the text is true)'. Christopher New (1997: 422) claims that it is determined by 'what the author's sentences mean [...] and what they logically or pragmatically imply in accordance with the conventions of fiction making prevailing at the time he composed the fiction'. And Stacie Friend (2011: 187) argues:

> that an utterance '*S*', occurring in the context of a discussion of a fiction F, is true if and only if F prescribes imagining that *S*. This will be either because '*S*' is to be understood as meaning, or getting across, something like 'According to F, *S*', or because it is to be understood as a continuation of the imaginative engagement authorized by F.

Probably the most influential account of truth in fiction, though, comes from Lewis (1978), according to which whether something is true in a fiction is, loosely speaking, determined by what is true at the nearest worlds where the fiction is *told as known fact*; the nearest worlds, that is, either to the actual world or to that set of worlds that represents the collective beliefs of the storyteller's community (for ease of exposition, we will focus on the former). In the rest of this paper we will primarily focus on a discussion of Lewis's view, but a number of the key claims we make may also be applicable (*mutatis mutandis*) to other accounts of fictional truth. In Section 1.1.4 we will outline Lewis's view, and its consequences, in more detail. First, though, we shift focus from fictional truth to fictional *indeterminacy*.

It is generally accepted that many (perhaps all) fictions are indeterminate in some respects. The most prominent kind of fictional indeterminacy involves a fairly straightforward kind of fictional incompleteness. There are many things about the inhabitants of fictional worlds that just aren't specified. Consider the famous case of Lady Macbeth and her children. It is clear from Lady Macbeth's claim 'I have given suck, and know how tender 'tis to love the babe that milks me' that she has at least one child, but it is almost universally agreed that there is no specific number of children which it would be correct to attribute to her. Indeed, Knights' (1933) intention in raising this question in his famous paper was to poke gentle fun at anyone inclined to think there *could* be a determinate answer to such inquiries. Similarly, while it is true in the fiction that many characters have full heads of hair, it would seem bizarre to believe that there is some fact of the matter about precisely how many hairs each of them has. Indeed, indeterminacy of this kind is so pervasive that some are inclined to maintain that in 'a fictional world it will normally be the case for each character *a* and some relevantly applicable predicate F that neither F*a* nor ~F*a* is true' (Heintz, 1979: 91). However, while cases of this kind have received the lion's share of attention in the literature, there are a number of other ways in which a more robust fictional *indeterminacy*, rather than this commonplace kind of incompleteness, might arise.

Most obviously, there are fictions that are indeterminate because they feature common-or-garden cases of vagueness, what Sorensen (1991: 66) calls 'vagueness within a story'. For example, we might be told that a character is 175 millimetres tall or that they have exactly *n* hairs on their head, but still judge it to be indeterminate whether they are tall or whether they are bald (in much the same way as we would when given this same information concerning some non-fictional individual).

Likewise, we may encounter a fiction such as Robert Anton Wilson's *Schrödinger's Cat Trilogy,* which presents us with various examples of indeterminacy in physics. How best to account for this kind of indeterminacy is, of course, a controversial matter. Someone who inclines towards the account of real-world indeterminacy we outline below will, most likely, also take it to be the best explanation for parallel kinds of indeterminacy as they arise within a fiction.[2] However, such cases hardly present an independent motivation for accepting our view. Such a motivation can, however, be found by investigating a third kind of fictional indeterminacy.

Consider the following questions. Is the governess in James's *The Turn of the Screw* haunted by supernatural apparitions or merely by symptoms of her own mental instability? Which of the two apparent realities in Smith's 'In the Imagicon' is the real world? Is the eponymous Babadook in the recent film a real monster or merely a representation of Amelia's grief? It seems to many (ourselves included) as if there are no determinate answers to these questions. We will, however, argue below that there is an important difference between these examples and the kind of indeterminacy as incompleteness discussed above. This difference is often overlooked in discussions of fictional truth.[3] In the Lady Macbeth case it seems entirely incidental to the narrative that there is no determinate fact of the matter concerning the number of her progeny. By contrast, an alternative version of 'In the Imagicon' that revealed in the final sentence which of the two worlds was the real one would, *ceteris paribus*, be a significantly inferior story. Just as a close parallel to *The Turn of the Screw* which took the form of an unambiguous ghost story would lack much of what made the original so engaging. The kind of indeterminacy we are focussing on in such works is not merely incidental but, rather, plays an important role in their artistic success.

We will argue in Section 3.3 that, for someone broadly in sympathy with Lewis's project, the best way to account for indeterminacy of this kind requires treating it as fundamentally different from the kind of indeterminacy as incompleteness that has been the main focus of discussion in literature. Before doing so, though, we will need to present Lewis's view on truth, and indeterminacy, in fiction in rather more depth (Section 1.1.4), as well as showing why we believe it is incapable of properly accounting for indeterminacy of this kind (Section 2.2). But first, we return to outline the account of metaphysical indeterminacy that provides our starting point.

1.1.3 The Barnes-Williams model for metaphysical indeterminacy?

We can now spell out in more detail the BW account of metaphysical indeterminacy. According to this influential view, 'there is the One True actualized world – it is just indeterminate which world this is' (2011: 131ff). In other words, there really is one actual world, but which one that is is itself indeterminate. Metaphysical indeterminacy is a matter of the indeterminacy of *which* world is the One True world. Formally, this means that instead of evaluating propositions at a specific (maximal, sharply defined) world, we evaluate them on sets of worlds.

In the BW model, the candidate 'precisifications of the world' are ersatz possible worlds: It is indeterminate whether p if some of the ersatz worlds that are candidates for representing actuality are p worlds, and some are ~p worlds. To be a candidate for representing actuality is to not determinately fail to represent actuality. (Thus, the BW model is not reductive in an important sense, but that is not our focus here.)

BW claim that this theory is fully bivalent and that it is committed to the principle of excluded middle. It is committed to the former because there is One True world (even though it is indeterminate which candidate it is), and for every possible world that is a candidate, for any proposition p, either p is true in that world or p is false in that world. And the theory is committed to the principle of excluded middle, because for any proposition p, (p ∨ ~p) is definitely true.

The supervaluational character of the models, and the connection to possible worlds, makes indeterminacy analogous to modality: Typically it will not be definite that p and not definite that ~p; thus indeterminacy is to definiteness as contingency is to necessity. And just as a disjunction will often be necessary while its disjuncts are contingent, so 'the university is in the city centre' and 'the university is in the suburbs' may both be indeterminate while 'the university is either in the city centre or in the suburbs' is determinately true.

The comparison with possible worlds is suggestive for our later proposal. To foreshadow our argument, we will claim that in order to account for deep indeterminacy we need a treatment analogous to that of modal logic using less-than-maximal possibilities instead of the more familiar possible worlds.

1.1.4 A Lewisian approach to indeterminacy in fiction

Having given some detail on the BW model for metaphysical indeterminacy, we can likewise explain more fully the Lewisian approach to fictional indeterminacy. Lewis (1978: 42) proposes the following influential account of fictional truth:

> A sentence of the form 'In the fiction f, φ' is non-vacuously true iff some world where f is told as known fact and φ is true differs less from our actual world, on balance, than does any world where f is told as known fact and φ is not true. It is vacuously true iff there are no possible worlds where f is told as known fact.

This account is relatively simple but has a great deal of explanatory power. In particular, it allows us to easily explain why many things are true within a particular fiction, even when not explicitly presented as being so in the fiction itself. It is true, for example, that the characters who inhabit most of our everyday fictions are ordinary human beings, because the nearest worlds where those stories are told as known fact are all ones where the individuals involved are human beings rather than, say, shockingly advanced robots.

Lewis's account of fictional truth also seems very well-placed to deal with certain kinds of fictional indeterminacy. As part of his analysis of truth in fiction, Lewis (ibid.) raises questions such as, 'Is the world of Sherlock Holmes a world where Holmes has an even or an odd number of hairs on his head at the moment when he first meets Watson? What is Inspector Lestrade's blood type?' and concludes that it would be absurd to suppose that such questions have answers. This is not, however, because there is any world in which the Holmes stories are told as known fact where these questions have no answers. Rather, such questions lack a correct answer because truth in a fiction is determined not by what is true in some single possible world, but by what is true in a *set* of worlds. The 'worlds of Sherlock Holmes are plural, and the questions have different answers at different ones' (ibid.). Those things that are true in every one of these worlds – that Holmes is a detective, that he is a human being, that he has never met Jeff Bridges and so forth – are true in the fiction. Those things that are false in every one of these worlds – that Holmes is a silicon-based lifeform, that he owns a mobile phone, that he was once married to Carly Simon and so on – are false in the fiction. However, what is true in some of these worlds and false in others is neither true nor false in the fiction, and any answer to the silly questions Lewis lists would fall into this category.[4] And, of course, what applies in the Holmes stories applies to the other cases of indeterminacy as incompleteness we have listed above. The claim that Lady Macbeth has exactly three children is true in some worlds where Macbeth is told as known fact but false in others. As such, it is neither true in the fiction that she has precisely three children nor true in the fiction that she does not.

Further, the Lewisian seems to have ample resources available to deal with the phenomenon of vagueness within a story. Whatever account we give of vagueness in the actual world, this account will, presumably, also be available to explain cases of vagueness in the nearest world where the fiction in question is told as known fact. We will, however, go on to argue (in Section 2.2) that the Lewisian account is unable to account for the third kind of fictional indeterminacy we have highlighted above.

2 The challenges of deep indeterminacy

Although we have sympathy with both the BW model of metaphysical indeterminacy and with the Lewisian model of fictional indeterminacy, they both suffer from a structurally similar problem. In both cases, the models are unable to adequately capture the full range of cases that fall under the relevant notion of indeterminacy. Further, they fail to do so in the same way.

There are versions of both metaphysical and fictional indeterminacy that are particularly resistant to domestication. This is what we call 'deep' indeterminacy. A form of indeterminacy is deep when it comes from the nature of the domain under discussion. Or, to put it differently, deep indeterminacy is the sort of indeterminacy that is built into the theory, rather than being a corollary of that theory. To make this clearer, consider the two cases we have before us. Deep metaphysical indeterminacy is non-representational indeterminacy that arises from the theoretical structure of accounts of the world. The particular instance we will discuss shortly relates to certain interpretations of quantum mechanics. On these interpretations, it is part of the nature of the quantum theory that worldly indeterminacy occurs. This is cashed out by mathematical constraints on possibility. Deep fictional indeterminacy is indeterminacy in fiction where the structure of the fiction itself is the source of the indeterminacy. It is, in a sense we will explicate further below, part of the nature of the fiction that there is indeterminacy. This can be cashed out by appeal to identity constraints on works of fiction: to be that very story is to contain indeterminacy of this sort.

Deep indeterminacy has the following feature: it is hard to supervaluate away. This is because constructing alternative candidates over which to supervaluate is complicated by the constraint that these candidates are both determinate and *genuine* candidates. How can fully determinate and complete accounts of the world be real candidates for being a deeply indeterminate world? In both the metaphysical and the fictional cases, it isn't possible to capture the indeterminacy in question within the models discussed.

Our answer is to adjust these models to give up on the completeness of the candidates; rather than dealing with possible worlds, we deal with smaller possibilities. We will also require controversial claims about the relationship between truths of propositions in different, mereologically related possibilities. But first, we will explain why there is reason to believe in such *deep* indeterminacy (both metaphysical and fictional) and spell out the inadequacy of sets of complete, determinate candidates as a means of accurately representing this.

2.1 Deep metaphysical indeterminacy

The possibility of deep indeterminacy, and the inadequacy of the BW account in capturing it, becomes apparent when we consider the way in which quantum mechanics is supposed to provide a motivating example for the metaphysician's project of theorising about metaphysical indeterminacy.

If quantum mechanics is to serve as an example of metaphysical indeterminacy, the obvious idea is to interpret Heisenberg's Uncertainty Principle as involving indeterminacy: If something has a definite position, then it lacks a definite momentum; its momentum is indeterminate. Likewise, perhaps it is indeterminate whether Schrödinger's cat is alive before the box is opened. So we have indeterminacy, and, importantly, its treatment has the same kind of 'supervaluational' structure as linguistic indeterminacy. In the appropriate state, it is definite that the cat is alive or the cat is dead, but not definite that the cat is alive or definite that the cat is dead; it is indeterminate whether the cat is alive.

To interpret the uncertainty principle in this way is not unusual, but requires very particular interpretative assumptions. We do not go into these details here – we propose to take for granted whatever assumptions one needs to get an example of metaphysical indeterminacy; we note just that those assumptions are controversial, but also commonly made.

There are certainly interpretations of quantum mechanics that, *prima facie*, involve genuine indeterminacy (although no concrete metaphysical account of that indeterminacy is part of the interpretation). The earlier versions of the GRW theory are likely candidates. One might worry, however, that in more recent interpretations, indeterminacy disappears along with the measurement problem. For example, in the modern Everett interpretation, the underlying ontology, from which the branches emerge, is perfectly determinate (whether the branches themselves are is another matter, but the underlying ontology is the level at which one would naturally judge the question of whether there is metaphysical indeterminacy). In the modern, 'flashy' GRW theory, the distribution of flashes in spacetime is also perfectly determinate. If one subscribes to the view that, one way or another, what is fundamental is the wave function evolving in configuration space, then, yet again, the fundamental ontology is perfectly determinate. Perhaps there is a sense in which derivative ontology that does include indeterminacy still falls under what is intended by 'metaphysical indeterminacy', but this is much less clear. So one might take the view that, initial appearances notwithstanding, properly interpreted quantum mechanics does not involve any metaphysical indeterminacy either.

Again, then, we register caution but proceed with the consensus (of Wilson (2013), Williams (2008) and Bokulich (2014), for example) that there is at least motivation from physics for seeking an account of metaphysical indeterminacy. An important point here is that there is good reason for thinking that indeterminacy in quantum mechanics is reliably worldly, not merely representational. The basic reason for thinking that the theory involves indeterminacy is because it does not give precise values for all observables (for example, no value for position in many states, and definitely not both a value for position and a value for momentum in any state at all). The reason for thinking that quantum mechanics involves metaphysical indeterminacy is that it does not even make mathematical sense to give precise values for all observables. This is the gist of the various results that culminate in the Kochen-Specker theorem and its subsequent refinements.

The details of this are unimportant for our purposes. What does matter is that there is substantial reason for thinking that there is metaphysical indeterminacy involved.

What really matters in the present context is that this very reason for thinking that quantum mechanics involves metaphysical indeterminacy presents a problem for many of the models whose authors invoke it as motivation. With Skow,[5] we present the challenge as an argument against the BW model, but suspect that it may also cause trouble for other models. The problem is simply that BW model indeterminacy using ersatz possible worlds: to be indeterminate is to hold in some but not all of the ersatz worlds that are candidates for representing actuality. Worlds by their nature are maximal in the sense of settling everything; thus in every world every observable gets some value. But the essence of the Kochen-Specker result is that it is impossible for every observable to have a value at once. Thus, this kind of 'deep' indeterminacy cannot be modelled by the BW framework.

There are a number of ways in which one might respond to this argument. For a start, if one begins with the vector space formalism and tries to read it as involving metaphysical indeterminacy, there are all of the interpretive assumptions that go into the particular interpretation of the formalism required (the eigenstate-eigenvalue link, and so on). These assumptions then have to be further strengthened (for example, by assuming that they have to hold at all of BW's ersatz worlds) in order to generate the problem for the models of indeterminacy. If, on the other hand, one thinks about things from the point of view of the specific proposals for solving the measurement problem, then it is not clear that they involve indeterminacy at all, *a fortiori* not deep indeterminacy. (This response is dialectically unattractive for the advocates of metaphysical indeterminacy, of course, since it removes not just the obstacle to its modelling but also its motivation.)

With this in mind, we propose to treat the objection as a plausibility argument, irrespective of its cogency from the point of view of the philosophy of physics. That is, we will not follow Skow in claiming that the Kochen-Specker result shows that deep indeterminacy is clearly possible, exactly, but just that it is at least important to explore how theories of metaphysical indeterminacy should handle it. In methodological terms related to the now familiar science-based objections to armchair metaphysics, the point is just to avoid leaving unnecessary hostages to fortune. In terms less influenced by the naturalistic objections, the point might be that even if one is completely unswayed by the scientific point, there is no reason from the armchair, now that it has been pointed out, why deep indeterminacy should be impossible. It sounds, on the surface at least, like it should be possible if any kind of metaphysical indeterminacy is, and so should be accounted for by an acceptable theory.

What we have here, then, is one motivation for exploring whether one can model deep indeterminacy in something like the BW framework.

2.2 Deep fictional indeterminacy

The second motivation likewise comes from deeper consideration of the kinds of fictional indeterminacy that are possible. We have seen that Lewis's account of fiction can provide a plausible explanation for certain varieties of fictional indeterminacy and, in particular, that he offers an attractive account of the kind of indeterminacy which arises owing to fictional incompleteness. Our third kind of indeterminacy concerning, for example, which of the two worlds in 'In the Imagicon' is real is often treated, when it is discussed at all, as merely a particularly interesting variety of this kind of indeterminacy. The only relevant difference is that the incompleteness is intentional rather than merely accidental. Matravers (1997: 425), for example, propounds something like this view when he claims that 'in *The Turn of the Screw,* Henry James toys with us by not providing conclusive evidence for choosing one of two worlds'.[6] That is, James deliberately provides insufficient information for us to judge whether *The Turn of the Screw* is a ghost story, just as Conan Doyle happens not to tell us precisely how many hairs Holmes has on his head at that fateful moment.

We think, however, that the indeterminacy in such cases is importantly different from the standard kind discussed above. It is not merely a matter of what the storyteller neglects to specify but, rather, it is essential to the nature and value of the work that no such specification is provided. Of course, there is some controversy with respect to precisely which works exemplify indeterminacy of this kind. There are, for example, those (such as Currie (1990: 182)) who are inclined to regard *The Turn of the Screw* as definitively a ghost story, and others who follow Kenton (1924) in subscribing to a straightforwardly psychoanalytic reading. However, the point remains that indeterminacy of this kind clearly exists in some fictions, and it is important for a full account of fiction, and fictional truth, to accommodate such fictive deep indeterminacy. As it stands, however, we believe that Lewis's account is unable to do so.

To see why, let's return to the contrast with the Lady Macbeth case discussed above. There are various worlds where *Macbeth* is told as known fact in which it is perfectly determinate that Lady Macbeth has precisely three children, others where it is determinate that she has precisely four and so forth. As such, while there may be some things that are indeterminate within the fiction (because they are true in some *Macbeth* worlds and false in others), each of these individual worlds is perfectly determinate. Or, rather more precisely, if such worlds are indeterminate, this will only be so because of features external to the debate over truth in fiction: standard cases of vagueness, indeterminacy in physics and so forth (when discussing whether fictional worlds are determinate below we will, for the most part, bracket such considerations).

In contrast, we maintain that there is no fully determinate possible world where stories such as *The Babadook* are told as known fact. There are, of course, worlds where certain determinate stories very much like *The Babadook* are told as known fact. Stories that determinately feature a real monster are told as known fact in some worlds, and stories where such supernatural entities are determinately absent are told in others. However, the crucial difference is that none of these are really tellings *of The Babadook*. Rather, they are tellings of much less complex and interesting stories that feature many of the same plot elements as *The Babadook*. *The Babadook* – and *mutatis mutandis* the *Turn of the Screw*, 'In the Imagicon' and so forth – is a successful work precisely because it is indeterminate with respect to the existence of the titular monster and the nature of the peril which the protagonists face. Any story which lacked these features, as either of the determinate fictions we have considered must, would (all else being equal) be inferior to *The Babadook* in certain key respects and could not, therefore, be the same story. As an extreme example of this phenomenon, consider the 21st Century Monads song 'Mr Determinable'.[7] This song tells the story of a curious individual who possesses a number of determinable properties while failing to instantiate *any* determinate of those properties. We are told that he has 'an occupation', 'an animal companion' and 'a number of children', but the narrative of the song makes it clear that there is no determinate answer to the question of what kind of animal companion, or how many children, he has. Again, there is no determinate world (and, *a fortiori*, no *set* of worlds) where such a story could be told as known fact. Any fully determinate version of the song would simply miss the point. Or so we claim. We will have more to say about why we favour an interpretation of this kind in Section 3.3. First, though, it is worth highlighting precisely why we take these cases to cause such problems for the standard Lewisian view.

It is a relatively simple matter to see why our account is in tension with the letter of Lewis's view. Lewis (1978: 43) explicitly rejects the claim that any individual world is itself indeterminate, claiming that

> I do not know what to make of an indeterminate world, unless I regard it as a superposition of all possible ways of resolving the indeterminacy – or, in plainer language, as a set of determinate worlds that differ in the respects in question.

As such, while Lewis's account can easily accommodate the kind of indeterminacy present in the Holmes case, we believe that it cannot adequately model the deeper kind of indeterminacy which arises in stories such as 'In the Imagicon'. This is because, in our view, there is no determinate world (and so no set of such worlds) in which these stories are told as known fact. Given this, it would seem that such stories fall under the second part of Lewis's analysis, which maintains that claims about truth in a fiction are vacuously true 'iff there are no possible worlds where [the fiction] is told as known fact' (ibid.: 42). This approach is clearly problematic, though, since there are many claims concerning such works that manifestly *aren't* true. The governess is not able to fly under her own power, she is not haunted by pink elephants and so forth.

It could, perhaps, be suggested that we could save a broadly Lewisian analysis by appealing (as some, such as Heintz (1979) have already suggested) not to ask what is true at a set of determinate worlds but, rather, to what is true at a single, indeterminate world. According to this suggestion, all indeterminacy can be dealt with this way. So, the world at which Macbeth is told is indeterminate with respect to the number of Lady Macbeth's children, the world of *The Babadook* is indeterminate concerning the existence of monsters and so on. (Of course, Lewis himself would not countenance such an appeal to indeterminate worlds as the quote above, and his arguments [Lewis, 1993], make clear, but we will assume that our reader is at least somewhat sympathetic to the idea of ontic indeterminacy.) We think that account has a number of advantages. Indeed, it parallels our own preferred solution in a number of respects. Our main concern, though, is that such an account would again blur the distinction between two importantly different kinds of fictional indeterminacy. In this account, the worlds where *Macbeth* is told as known fact would contain a woman who has an indeterminate number of children, just as the titular Mr Determinable does in the Monads' song. Such an account would, therefore, once again leave us without the resources to explain what is interesting and amusing about the latter. Mr Determinable would be no different than the standard kinds of fictional characters who appear in Shakespearean plays, Dickens novels, Hollywood blockbusters and the like. Of course, if it transpires that there is no plausible Lewisian account that is able to respect this distinction, then this would give us reason either to reject Lewis's view entirely or to explain away the apparent difference between the cases of indeterminacy we have surveyed above. We will, however, argue below that this is not the case and that a broadly Lewisian count can be constructed, which is able to accommodate the important difference between these two kinds of fictional indeterminacy.

3 Modelling deep indeterminacy

3.1 Using situations

The problem that deep indeterminacy poses for accounts of metaphysical indeterminacy can be solved by using *situations*, parts of possible worlds rather than whole worlds, to do the modelling.[8] Situations can be most easily understood as parts of possible worlds. However, for the situation theorist this gets things the wrong way round: they will take situations as their theoretic primitive and define possible worlds as special sorts of situation. The exact nature of situations is a topic of some debate in situation theory.[9] But whatever situations are, they are inherently partial; they do not tell us everything.[10] For instance, the situation encompassing only the Second World War does not give us any information about the First World War, or future Martian settlers, or events happening on a distant planet in 1945. The claim of situation semantics is that it is with respect to situations that expressions or sentences are evaluated.

Evaluating with respect to situations is claimed to have a number of advantages. For just one of them, and to see the semantics at work, consider the following example from Barwise and Etchemendy (1987).[11] Suppose that Emily is playing cards and has the Three of Clubs.[12] Suppose, further, that Claire is also playing cards elsewhere in the city and also has the Three of Clubs. If I am watching Emily play and mistake her for Claire, I might utter: 'Claire has the Three of Clubs.' It might seem as though what I have said isn't true. This is captured by the situation semanticist: the particular situation I'm talking about doesn't contain Claire, so doesn't include Claire having the Three of Clubs. In general, only the contents of the particular situation can make a statement true. It matters, then, which particular situation is being picked out by a statement. This sort of contextual element is exploited by situation semanticists to capture the intuition that the statement above is not true. Situation semantics has an advantage here over possible world semantics. Possible world semantics is coarser-grained because the entities it considers (worlds) are larger than situations. Possible world semantics has my utterance true, as Claire does indeed have the Three of Clubs in the world in which I utter the statement.

This, then, can be seen as one example where situation semantics gives a distinctive interpretation. There are, of course, different formalisations of situation semantics and a number of open questions in situation theory. We shall settle only those that are necessary for the project at hand. The first to mention is that we have a choice about the truth-value of statements like 'Claire has the Three of Clubs' in scenarios such as the one described above. The proposition expressed is not true in a situation that doesn't contain Claire, this much is clear. But is it false? Or is it neither true nor false? We favour the latter view, as the situation seems to simply not tell us anything about Claire and her possession of the Three of Clubs. It is not that she is in the situation and doesn't have that card, but rather that the situation, being partial, leaves open whether she has it or not. Thus, the situation semantics we are using is one according to which propositions can lack truth-values. This should not be a surprise: the partial nature of situations lends itself to a view according to which situations do not determine a truth-value for every proposition.

Two accepted parts of situation theory can now be added to the picture. They are the following:

(i) Situations can be parts of one another, and
(ii) Evaluation of the truth-value of propositions is relative to a situation.

This gives us a relation between situations and an account of truth that is situation-relative.

The parthood relations between situations together with the situation-relative truth-values of propositions invite questions about when the truth-values of propositions agree across situations related in different ways.[13] Our central question is this: if some proposition p is true in some situation s, is p thereby true in all situations which have s as a part?

Our answer is no. This issue is known as the question of persistence, and is one on which situation semanticists are divided.[14] An initial reason to doubt that propositions always persist to their extensions comes from certain quantified statements. The proposition expressed by 'all the students passed the test' can be true in some small situation involving, say, only one class taking one test. But this proposition will not be true in some larger situations containing the former, for instance a situation containing all tests in a region that year. Of course, there are ways to maintain persistence, perhaps by positing implicit domain restriction in cases like the above, and replies are available to such moves too. There are other cases, though, where denial of universal persistence for 'atomic' propositions, which do not contain quantifiers, seems metaphysically productive (Pickup [2016] indicates a solution to the Ship of Theseus puzzle using similar machinery). At any rate, the denial of universal persistence is key to the model we present. So, for the sake of this solution, we will take it that propositions true in some situation are not always true in any situation that has it as a part. But, of course, propositions do often persist. Though it isn't necessary for what comes later, we think that a restricted version of persistence akin to, but not identical to, that proposed in Pickup (2016) can be endorsed:

> Restricted persistence: if p is true in s and $s \leq s*$ then p is true in s∗ unless there is an $s' \leq s*$ such that some q is true in s' and q is incompatible with p

Now we shall show how the use of situations, along with a denial of universal persistence, can give us the right structure to express the claims of deep indeterminacy in a way that allows the cases from physics and fiction we have outlined.

3.2 Modelling deep indeterminacy in physics

As noted, the trouble with the BW model is that it requires complete precisifications (in the form of entire possible worlds) as the candidates between which world is not settled. In short, as mentioned above, situations can be used as precisifications of incomplete things, and denying universal persistence will rule out problematic inferences. What it means for something to be metaphysically indeterminate is for a number of different, incompatible situations to be real candidates for the way reality is. We can maintain that these different candidate situations are all actual. But, as we shall see below, what is true in them doesn't necessarily persist to the actual world, since this would lead to contradiction and, for the quantum mechanical case, mathematical impossibility.

To spell this out for quantum mechanical metaphysical indeterminacy, consider a number of distinct situations. The first, s_1, contains certain information about the properties of a quantum system. It might, for instance, have a particle spin-up in the x direction. Let this information be captured by a proposition p. Thus, in s_1, p is true. The second situation, s_2, is one in which the quantum system has contrary properties, e.g. the particle has opposite spin in the same direction. Thus, in s_2, the proposition ~p is true.[15] So far, nothing differs from the BW proposal. But here's the distinctive advantage of situations: not all of the properties of the quantum system need to be precise in s_1, which need, for example, include nothing about spin of that particle in the y-direction. Now take two further situations, s_3 and s_4. In s_3, some proposition q, corresponding to some value for spin in the y-direction, is true, and ~q is true in s_4.[16] In a scenario involving deep indeterminacy, it might be that no situation can contain determinate truth-values for both p and q.[17] But, we can hold q to be neither true nor false in s_1 and s_2. And ~q will likewise lack a truth-value in s_1 and s_2. Similarly, both p and ~p will fail to have a truth-value in s_3 and s_4. Thus in none of s_1-s_4 is there conflict with deep indeterminacy.

Let's recap. We have four situations and in each one of a pair of propositions is true (and the other false), while a second pair receives no truth-value. These situations are the candidate precisifications that give rise to the indeterminacy of reality with respect to the propositions. It is unsettled which propositions are true. So far, so good. But there's a further question to ask here. What are the parthood relations that s_1-s_4 stand in? Obviously none of them are parts of one another. But we can construct larger situations out of them (in standard situation theory situations obey unrestricted and unique fusion). Consider, in particular, whether the situation S that is the whole world contains as parts all of s_1-s_4. On the version of the situation-theoretic approach we are outlining here, S does indeed have all of s_1-s_4 as parts: the world contains all of its candidate precisifications. But doesn't this mean that the world itself is contradictory, as it contains parts that determinately disagree about the truth of some propositions? Now we see the benefit of the denial of universal persistence. For if we accepted universal persistence, we would be required to hold that anything true in a part of S is thereby true in S. Thus p and q would both get determinate truth-values, in violation of deep indeterminacy, and furthermore p and ~p would both be true, in violation of non-contradiction. Instead, we maintain that these propositions, though true in parts of S, are not true in S. They fail to persist from the sub-situations to the larger situation: they lose truth-value. The restricted persistence condition above gives this result. In this way, the world can contain parts that fundamentally disagree but not be contradictory.

To spell out the view: we propose to replace the original BW structure with a new structure. BW evaluated propositions with respect to sets of ersatz possible worlds. These possible worlds are the candidate precisifications of reality.

For BW, it is indeterminate whether p if some of the ersatz worlds that are candidates for representing actuality are p-worlds, and some are ~p-worlds. To be a candidate for representing actuality is to not determinately fail to represent actuality.

Our structure evaluates propositions with respect to sets of (sharply defined but typically local and incomplete) situations. These situations are the candidate precisifications of parts of reality, and are all parts of the maximal actual-world situation S. It is indeterminate whether p if some of the situations that are parts of S are situations in which p is true and some of the situations that are parts of S are situations in which p is false. Truths in parts of situations persist to whole situations unless they conflict with truths in another part of the whole situation. If they do conflict, they are indeterminate in the whole situation.

3.3 Modelling deep indeterminacy in fiction

In this section, we present our own modified Lewisian account of fictional indeterminacy. Before doing so, though, it is worth noting that the kind of indeterminacy we are concerned with – as with the kind of deep physical indeterminacy we discussed above – cannot be modelled using a standard version of the BW account. Recall that each of the possible worlds they postulate ascribes fully determinate truth-values to every proposition. The source of metaphysical indeterminacy lies not in the worlds themselves but, rather, in its being indeterminate which of them is actual. As such there will still be no world where a story like *The Babadook* is told as known fact. Instead, there will only be worlds where, for example, some less interesting monster story is told. Given this, the Lewisian who adopts the BW account of indeterminacy will face precisely the same problem encountered by Lewis's original view.[18] On our account, by contrast, there will be possible situations in which stories of this kind, and not merely determinate parts or variations of them, are told as known fact. A narrator in one of these situations can truly tell a story that is indeterminate between the different interpretations of *The Babadook* because the situation they inhabit is itself indeterminate in these respects.

To see how this might come about, start by considering two possible situations. In one of these (B_1) a close cousin of *The Babadook* story, which features a real monster, is told as known fact, and in the other (B_2) a story where there is no monster is told as known fact. Now, consider a third situation (B_3) which combines the two. It is in B_3, we maintain, that someone will be able to tell *The Babadook* itself as known fact. While it will be determinate with respect to B_1 that there is a monster and determinate with respect to B_2 that there is not, there will be no determinate fact of the matter regarding the monster's existence in B_3. Recall that, on our model, all of those things which are true in B_1 (and likewise B_2) will be true in B_3 unless they conflict with what is true in some other situation which is also part of B_3. So it will be straightforwardly true in B_3 that Amelia has a son called Samuel, that they live in Australia and so forth. Conversely, it will be straightforwardly false that they reside in Peru and that they lead happy, carefree lives. However, it will not be determinately true in B_3 that there is a monster (since this conflicts with what is the case in B_2) nor that there isn't (since this conflicts with what is the case in B_1).[19] Rather, the truth-value of the claim 'there is a monster' will be indeterminate.

What is true in *The Babadook* will, therefore, be what is true in each of the situations such as B_3 where the story itself is told as known fact. There will, however, be two ways for a proposition to be indeterminate with respect to such a fiction. The first (shallow fictional indeterminacy) arises when a claim is true in some Babadook worlds and false at others. The second (deep fictional indeterminacy) occurs when the truth-value of a proposition is itself indeterminate in each of the Babadook worlds. In the Holmes stories, by contrast, we only have indeterminacy of the first (shallower) kind. Thus we have a clear Lewisian explanation for the important, and neglected, difference between these different kinds of fictional determinacy.

4 Conclusion

We have explored two areas that motivate theorising about indeterminacy – in fiction and in physics. Moreover, when we consider in more detail the *kinds* of indeterminacy involved, they both appear to feature something that one might initially not suspect: *deep* indeterminacy. This is significant because the natural ways of thinking about indeterminacy in both domains appear to be unable to handle deep indeterminacy. At this point, in other cases involving friction between philosophy and other domains of enquiry, various moves are familiar in the philosophical literature, and all might be made here.

For example, a conservatively inclined philosopher might dig in and declare the phenomenon in question impossible. In the case of deep metaphysical indeterminacy they might see the impossibility of deep indeterminacy – or of metaphysical indeterminacy at all – as a constraint on the proper interpretation of quantum mechanics. If, say, a version of GRW on offer involves indeterminacy in the world, then reject that version of GRW. In the case of deep fictional indeterminacy, the conservative line would be to declare that an author, no matter what their intention, simply *cannot* produce a fiction that is indeterminate in the relevant sense. They can make one that is *incomplete*, or one in which a fictional narrator *thinks* that there is indeterminacy, but not one that is genuinely indeterminate (see Hanley, 2004: 120 for parallel moves concerning certain kinds of impossible fiction).

Alternatively, a 'naturalistically' inclined metaphysician would see the constraint the other way round: conservatism is insufficiently deferential to the source of the phenomena. In the case of the BW model, perhaps the philosophers' fetish for nice, complete *possible worlds* is at fault. In the fiction case, again it is the Lewisian orthodoxy and the machinery of worlds that is seen to be inadequate to the task.

In adapting the models of indeterminacy in question, we have accepted something like the latter challenge: rather than dig in, the metaphysician ought to adapt. But unlike the more adventurous naturalistic metaphysician, we prefer to emphasise the naturalness with which orthodox analytic metaphysics provides the answers – the move to situations is nothing like the wholesale rejection of this kind of framework advocated by its detractors, such as Ladyman and Ross (2007). On the contrary, mainstream analytic metaphysics is a much richer source of tools, as is reflected in the 'Viking' approach of French and McKenzie (2012). And in the context of the present volume, the important point is this: the richness of those tools only becomes apparent when one considers the full range of their application; problems drawn from areas as apparently disparate as the philosophy of art and philosophy of science motivate common solutions and provide mutual insight.

Notes

1 For an explanation of the key notion of a 'fictional author', see Currie (1990: 75–80).

2 There may, of course, be some exceptions to this, such as philosophical fictions where we are explicitly told that some other account of indeterminacy holds.

3 Though, of course, *The Turn of the Screw* is itself frequently discussed in such debates (see, e.g., Currie (1990: 66–7), Matravers (1997: 425) and Stecker (1995: 16).

4 Well, not quite any answer. It is, for example, determinately true that Lestrade's blood type is one of those possessed by some relatively healthy human beings in the actual world.

5 The term 'Deep Indeterminacy' is from Skow (2010); we are using it in a slightly looser sense than Skow here.

6 And Stecker (1994: 201) makes similar remarks.

7 A full version of the song's lyrics can be accessed here http://the21stcenturymonads.net/lyrics/MrDeterminable.html.

8 What follows uses the formal machinery of situation semantics. There are other semantics akin to situation semantics which might be conducive to a similar approach. In particular, Fine's Truthmaker Semantics might be a fruitful alternative way to express the metaphysical claims here advanced. What will be crucial is the partial nature of the entities under consideration and the denial of monotonicity that occurs below relating to persistence.

9 Barwise (1988) lists a series of choice points for the nature of situations and their behaviour.

10 One might define possible worlds as special sorts of situations that are maximal in some sense. Such situations would not be partial in that sense. There might also be an unrestrictedly maximal situation that described everything that is possible (akin to the Lewisian pluriverse); this would not be partial in any sense. Whether there are possible worlds or a maximal element are open questions in situation theory, but the point here is simply that situations do not need to be maximal in any sense.

11 Kratzer (2011) describes some other areas within semantics where situations have been used productively, including anaphora and (possibly) the Liar paradox.

12 The example they use is in fact not ideal; by using a definite description it adds a level of complexity. But let's take it that by saying that someone has \`the Three of Clubs', we simply mean that they have a card with a certain property. The example would work at least as well with \`three clubs' rather than \`the Three of Clubs'.

13 One question is: if some proposition p is true in some situation s, is p thereby true in all parts of s? The answer is no, but this doesn't have an impact on the models we present. It should be obvious, however, that the partial nature of situations means that they don't settle the truth of everything true in their extensions.

14 For instance, Kratzer (1989) keeps hold of persistence, but Elbourne (2005) rejects it.
15 Depending on the exact semantics, this may be equivalent to the falsity of p.
16 There would also be further situations in which other propositions are true, in the quantum mechanical case, but two pairs suffice for the demonstration.
17 The Kochen-Specker theorem strictly applies only to more complex structures of observables, but the conceptual point is the same (recall that we only take the quantum-mechanical case as a motivating example).
18 There is also the difficulty of combining something like the BW account with a genuine, rather than ersatz, form of modal realism. We will not, however, pursue such concerns here.
19 Though it will be determinately true that there either is or isn't a monster, since this disjunction is true in both B_1 and B_2.

References

Akiba, K. (2004). 'Vagueness in the World'. *Noûs*, 38 (3), 407–29.

Barnes, E. and R. Cameron. (2009). 'The Open Future: Bivalence, Determinism and Ontology'. *Philosophical Studies*, 146 (2), 291–309.

Barnes, E. and R. Williams. (2011). 'A Theory of Metaphysical Indeterminacy'. *Oxford Studies in Metaphysics*, 6, 103–48.

Barwise, J. and Etchemendy, J. (1987). *The Liar: An Essay in Truth and Circularity*. New York: Oxford University Press.

Bernstein, S. (forthcoming) 'Causal and Moral Indeterminacy'. *Ratio*.

Bokulich, A. (2014). 'Metaphysical Indeterminacy, Properties, and Quantum Theory'. *Res Philosophica*, 91 (3), 449–75.

Currie, G. (1990). *The Nature of Fiction*. Cambridge: Cambridge University Press.

D'Alessandro, W. (forthcoming) 'Explicitism about Truth in Fiction'. *The British Journal of Aesthetics*.

Elbourne, P. (2005). *Situations and Individuals*. Cambridge, MA: MIT Press.

French, S. and K. McKenzie. (2012). 'Thinking Outside the Toolbox: Toward a more Productive Engagement between Metaphysics and Philosophy of Physics'. *European Journal of Analytic Philosophy*, 8 (1), 42–59.

Friend, S. (2011). 'The Great Beetle Debate: A Study in Imagining with Names'. *Philosophical Studies*, 153 (2), 183–211.

Hanley, R. (2004). 'As Good as It Gets: Lewis on Truth in Fiction'. *Australasian Journal of Philosophy*, 82 (1), 112–28.

Heintz, J. (1979). 'Reference and Inference in Fiction'. *Poetics*, 8 (1–2), 85–99.

Keefe, R. (2008). 'Vagueness: Supervaluationism'. *Philosophy Compass*, 3 (2), 315–24.

Kenton, E. (1924). 'Henry James to the Ruminant Reader: The Turn of the Screw'. *The Arts*, 6, 245–55.

Knights, L. C. (1933). *How Many Children Had Lady Macbeth?: An Essay in the Theory and Practice of Shakespeare Criticism*. Cambridge, MA: The Minority Press.

Kratzer, A. (1989). 'An Investigation of the Lumps of Thought'. *Linguistics and Philosophy*, 12, 607–53.

Kratzer, A. (2011). 'Situations in Natural Language Semantics'. *Stanford Encyclopedia of Philosophy*.

Ladyman, J. and D. Ross. (2007). *Every Thing Must Go*. Oxford: Oxford University Press.

Lewis, D. (1978). 'Truth in Fiction'. *American Philosophical Quarterly*, 15(1), 37–46.

Lewis, D. (1993). 'Many, but Almost One', in J. Bacon (ed.) *Ontology, Causality, and Mind: Essays in Honour of DM Armstrong*. Cambridge: Cambridge University Press, 23–42.

Matravers, D. (1997). 'Truth in Fiction: A Reply to New'. *The Journal of Aesthetics and Art Criticism*, 55 (4), 423–5.

New, C. (1997). 'A Note on Truth in Fiction'. *The Journal of Aesthetics and Art Criticism*, 55 (4), 421–3.

Pickup, M. (2016). 'A Situationalist Solution to the Ship of Theseus Puzzle'. *Erkenntnis*, 81 (5), 973–92.

Skow, B. (2010). 'Deep Metaphysical Indeterminacy'. *The Philosophical Quarterly*, 60 (241), 851–58.

Sorensen, R. A. (1991). 'Fictional Incompleteness as Vagueness'. *Erkenntnis*, 34 (1), 55–72.

Sorensen, R. (2001). *Vagueness and Contradiction*. Oxford: Oxford University Press.

Stecker, R. (1994). 'Art Interpretation'. *The Journal of Aesthetics and Art Criticism*, 52 (2), 193–206.

Stecker, R. (1995). 'Relativism about Interpretation'. *The Journal of Aesthetics and Art Criticism*, 53 (1), 14–18.

Williams, J. R. G. (2008). 'Ontic Vagueness and Metaphysical Indeterminacy'. *Philosophy Compass*, 3 (4), 763–788.

Williams, J. R. G. (2014). 'Nonclassical Minds and Indeterminate Survival'. *Philosophical Review*, 123 (4), 379–428.

Williamson, T. (2002). *Vagueness*. New York: Routledge.

Wilson, J. (2013). 'A Determinable-Based Account of Metaphysical Indeterminacy'. *Inquiry*, 56 (4), 359–85.

Woodward, R. (2011). 'Truth in Fiction'. *Philosophy Compass*, 6 (3), 158–67.

7 Some philosophical problems of music theory (and some music-theoretic problems of philosophy)

Dean Rickles

1 Introduction

Music has been investigated by philosophers for some time, especially (most recently) within the context of the metaphysics of music. Its ineliminable temporal aspects[1] as well as its ability to be multiply instantiated (via distinct performances or playings) renders it an ideal case study for matters ontological. Such studies have tended to restrict the focus to *works* of music, rather than general principles of music. For example, they might ask 'what *kind of thing* is Beethoven's Ninth Symphony?' This is an interesting question that has led to the application of many old viewpoints (nominalism, platonism, etc.), in addition to some novel positions linking to work originally carried out in modal metaphysics and the philosophy of time (and personal identity). Another area that has been investigated with interest by philosophers is the relationship between music and the emotions (and the puzzle of the expressiveness of music). However, as interesting as these topics are, this restriction leaves out much that might be of interest to philosophers of a different stripe, as well as leaving out aspects that might involve philosophy (of science) contributing something back to the study of music theory and musicology.

This chapter will shine a philosophical light (specifically: philosophy of science) on music theories and musicology (the more general study of music) and in particular the question of musical laws, models and theories.[2] To the best of my knowledge, philosophers of science have yet to include musicology or music theory among their objects of study.[3] This might be because musicology is in fact not to be considered a science. However, there are theories and models (especially of tonality and meter) that offer a good fit to the kinds of analysis carried out by philosophers of science, including various laws (or quasi-laws) and a central role played by abstract representation, often by the kinds of geometric spaces (such as configuration spaces, common in the physical and biological sciences[4]). We might note also that – historically, at least – musicology was certainly considered to be 'the science of music', and this remains part of most dictionary definitions (see L. Harap, 1937 – although he attempts to impose a criterion of 'rigour' instead of 'scientificity'). Indeed, some influential approaches to music theory were modelled directly after the natural sciences, with the concomitant idea that music theories involve the discovery of natural laws that underlie the actions of composers (and that are involved in the listening process). Hugo Riemann's approach was especially explicit (see S. Burnham, 1992).[5]

Here we focus primarily on meter and tonality, key aspects of musical structure which have several features (philosophically interesting features) in common due to their more subjective natures – that is, neither meter nor tonality are strictly speaking a part of the auditory signal (the stimulus) in which music is (at least partially) encoded: one and the same signal can be interpreted as having distinct meter and tonality by different observers (perhaps a direct analogue of underdetermination of theory by data). Meter and tonality have also received highly mathematical treatments, as mentioned above, and are closest to genuine scientific theories.

But, as we will see, we face an unavoidable problem: there exists a tension over what any such theories (and laws) might be referring to, and therefore over what music theory is a theory *of*. The problem is that there exists an ambiguity over whether a theory of music (and so the laws and symmetries) describes music *qua* objective structure (formal or otherwise) or *qua* subjective experience: objective musical structure versus experiential states of observers (phenomenology or the stuff of psychology and neuroscience).[6] Scruton distinguishes between 'the intentional object of a listener's experience' and 'the material organization of *sounds*' (1999).

I argue that Arthur Eddington's little known position of 'selective subjectivism' (as outlined in his *Philosophy of Physical Science*) is an appropriate response to this tension (it might in fact be one of the few sensible applications of Eddington's stance). Note that this is a tension that music theorists are themselves well aware of; it has usually been sidestepped by falling into one of two camps: either establish music theory as a theory of the mind and cognitive (perceptual) structure, or else entirely divorce music theory from the mental and cognitive structures that the theories generate (and/or are generated by).[7]

We begin in the next section with the fundamental materials of music theory, following which we consider some theoretical frameworks for dealing with these. We then develop the tension mentioned above (which already appears in the basic development of the musical materials and theories thereof), after which we proceed to show how selective subjectivism offers an appropriate resolution.

2 Elements of musical structure

First, let us distinguish between musicology and music theory. There are many overlaps, but musicology is certainly more general than music theory, including aspects of history, style, notation, canon and so on.[8] On the other hand,

I take music theory to be about purely musical *structure* (*pace* the ambiguity over whether this structure is objective or subjective). Meter and tonality form two central components of (Western) musical structure and are implicated in our expectations about the melodic and temporal organisation of musical works – other important components include harmony, rhythm, melody, timbre, tempo and pitch.[9]

Pitch is more directly observable than tonality or meter, and is at least an essential part of tonality perception. It is essentially the mind's way of representing the periodicity of sound waves. But it is relative pitch (the distance between different pitches) that is central to music perception (and musical structure); that is, one can change absolute pitch, leaving relative pitches invariant, and 'preserve the musical structure' in so doing.[10] This is an example of a musical symmetry. It seems that relative pitch is a basic feature of the human perceptual system. For example, even infants can recognize transposed melodies as *the same* (hence, the symmetry). However, melodic contour information (the movements of a melodic line up or down, for example) is easier to assess than specific interval information (although with the same deficits present as those experienced by sufferers of amusia, who are also unable to register these shifts) – untrained listeners, for example, sometimes have difficulty distinguishing major from minor intervals (with octaves being a notable exception, which there is universal ability to detect, even from a very young age, which is why octave-separated notes get the same name).[11]

Even at this relatively low level, in terms of the elements of musical structure, there is much cognitive processing going on. Pitch, after all, is not just frequency; it is linked to perception, and it *demands a subject*. Frequency does not. As one builds up to the concept of tonality (providing the sense of order and centeredness in a piece of music[12]), ever more cognitive load is involved. Thus, when Rameau declared, at the outset of his *Treatise of Harmony* (1984, p. 3), that 'Music is the science of sounds; therefore sound is the principal subject of music', he left a huge gap in understanding what sound is (himself leaving the problem of defining sound 'to physics'). But, further, the reason why the sounds have that particular musical structure and organization (and why the composer made it so) has much to do with the human mind. For Rameau, it was mathematical principles that were fundamental to this structure: the division of the string and the related harmonic series could explain the nature of tonal music. Rameau excepted, there are very few music theorists who view music theories as the study of raw physical sound signals. Most modern work on music theory acknowledges that music involves a certain amount of activity on the part of an observer. As Lerdahl and Jackendoff put it: 'Insofar as one wishes to ascribe some sort of "reality" to these kinds of [musical] structure, one must ultimately treat them as mental products imposed on or inferred from the physical signal' (1983, p. 2).

Including meter into our musical structure brings in even more 'mind stuff'. There is no sense in which meter 'exists in the physical signal' issued from a performance of music. As Danuta Mirka spells it out, '[T]he function of meter is to provide a cognitive matrix for rhythmic patterns in order to make their understanding possible' (2009, p. 22). Or, as Eric Clarke puts it, '[Meter is] a framework around which individual notes are organized, and through which they gain appropriate durational quantification' (1987, p. 228). In this sense, meter is somewhat like a spacetime metric, providing the background in which physical processes occur. In the case of music, meter is the background against which heard rhythm and 'rhythmic surface' are defined. However, absolute Newtonian space (in the sense of a rigid container for processes) does not seem to provide the appropriate analogy here. Though we do often, in fact, infer the meter from the observable rhythmic surface, there's a strange bootstrapping in this case, in which that very rhythmic surface is defined by the background meter. In this case, the cause/effect relation is unclear: the beats of an observable rhythmic structure generate a meter that shapes the pattern of those very same beats! It seems that a dynamical conception of meter more closely analogous to the metric in Einstein's theory of general relativity is more appropriate, in which the slogan 'matter tells space how to curve, and space tells matter how to move' sums up the embrace.[13]

However, such interesting parallels aside, what is important here is that meter, though a central piece of musical structure, is nowhere present in the sound signal entering the ear. As Martin Clayton puts it,

> Meter is more than a simple accentual pattern, and moreover it is not necessarily measurable or objectively demonstrable. On the contrary, meter depends for its existence on the agency of a human interpreter.
> (2000, p. 33)

That is, meter is disconnected from observable pulses – music theorists usually call the pattern of (strictly unheard) accents in meter 'beats' to distinguish them from the directly observed pulses of rhythm. Musically, meter is essential to keep a sense of structure and organisation as the pulses can drop out, shift pace or syncopate (fall off the pulse). This idea that a central component of music (and music theory) is created by the brain will be important to the application of selective subjectivism later on.

This provides some sense of the elements of musical structure (along with some of their philosophically salient quirks), which a theory of music will attempt to describe and/or explain. In a particular music, not all elements need be present. For example, African drumming music will have meter and rhythm present, and the drums will be tuned to some pitches, but harmony, tonic and melodic contours will, in general, not be present. Hence, it is difficult to provide models of the kind that one might give in philosophical discussions of spacetime theories.[14] But it is not impossible. After all, music is, at one level (capable of formalization) just organised sound: pitches or tones laid out in space (harmonically) and in time (rhythmically). Though not presented in quite this manner, many modern theories of music perform in just this way, specifying the basic elements and their possible relations, along with a kind of dynamic (given by constraints that may or may not issue from biological/psychological features of listeners).

3 Theories and models of musical structure

Music theories aim to describe different kinds of musical structure and how these different kinds of music organise the materials of music. One usually has to make a distinction between tonal and non-tonal (or atonal) music, since theoretical principles and explanations don't tend to generalise to all musics, and indeed it is seen as a virtue of most theories that they have enough specificity to distinguish tonal music from non-tonal music. The notion of a 'grand unified theory' of music that encompasses all types of music without alteration of some fundamental parameters (akin to denying some postulate of Euclidean geometry to generate non-Euclidean geometries) is liable to be rather trivial.[15]

In his *Explaining Tonality: Schenkerian Theory and Beyond*, Matthew Brown considers theories of tonality from the point of view of theories in other scientific domains. He draws attention to the fact that even the most basic of methodological questions that one might ask in other fields face problems in the case of tonality (and, I would add, meter). For example, the question of selection and rejection of particular theories is difficult on account of the polysemicity of the central term 'tonality'. Because of this, it is also difficult to know what such theories should be *about* and so what counts as success. There is even disparity over what a theory of tonality (and music theory) in general is supposed to achieve and how it is to be carried out, with focus on the purely musical structure ('internalist') or on the psycho-socio-cultural mileau ('externalist'). However, Brown *et al.* provide a list of desiderata for a theory of tonality (that can easily generalise to meter and other structural elements of music):

> [A] powerful theory of tonality should allow us to do more than merely separate the tonal sheep from the non-tonal goats; it should also enable us to predict how suitably qualified auditors might respond to features characteristic of tonal music. This list might include judgments about closure, completion, openness, stability, transition, goal-directedness, and many other related musical phenomena. The better the theory predicts such judgments, the more powerful it will be. Finally, we will say that a theory of tonality is fruitful if it also predicts (with relevant modifications) aural judgments about pieces that are not paradigmatically tonal. For example, it should explain the tonal tendencies of Debussy's music or Stravinsky's neo-classical works and indicate when and how the pieces deviate from common-practice conventions.
>
> (Brown *et al.* 1997, p. 157)

In this, we find a focus on the 'judgement' and experience of listeners. As we saw in the last section, this is entirely appropriate given the entangled nature of music theory in which the object of analysis is also partly subjective, such that the state of the analyst becomes part of the system analysed.[16] Lerdahl and Jackendoff quite explicitly take the goal of a theory of music to be '*a formal description of the musical intuitions of a listener who is experienced in a musical idiom*' (1983, p. 1).[17]

This kind of approach, involving a probing of the 'fine structure' of music, as inferred from a 'musical surface' harks back to the ideas of Heinrich Schenker.[18] This is a view that, although not strictly formalised as such, has at least an affinity with theories in the natural sciences. It attempts to uncover various levels (a hierarchy) of music structure. Those 'surface' phenomena that we directly perceive are related in various ways to deeper layers of organisation. In Schenkerian analysis one has, methodologically speaking, something that looks like the standard physicist's way of reducing a phenomenon to more fundamental units (though in this case the units are larger blocks). One extracts ever more extended progressions from the 'musical surface'.

Similarly, in considering 'syntactic theories of music', Marvin Minsky takes the aim of such theories to provide an answer to the question, 'Why do we like certain tunes?' (ibid., p. 336) – that is, he also indicates that music theory should concern itself with the experiences of listeners. He contrasts two answers, one involving *structural features* (encoding laws and rules of composition) and one involving the psychological fact that they match what we already know and have built expectations about (i.e. from hearing specific tunes during childhood). Minsky is not convinced that the former approach is viable, since he doesn't see that such rules and laws exist, and so he focuses on the psychological (developmental) perspective, through involving AI and simulations of the development of melodic sense and preferences. However, in their scheme, Lerdahl and Jackendoff, for one, extract a set of rules for assessing the well-formedness of musical structure. These involve four basic principles:

- Grouping Structure: the mind's method of grouping together musical events that sound like they belong together, into a linear stream.
- Metrical Structure: background beat structure, to which musical events are related.
- Timespan reduction: the selection of important musical events.
- Prolongational reduction: a global (Schenkerian) principle having to do with the analysis of tension and resolve, possibly over very long timescales (as related to shifts in pitch and rhythm).[19]

In addition to these, there are also 'Preference Rules' for each well-formedness rule. These are based on the 'relatively unchanging cognitive foundations of the musical mind' (2001, p. vii). In more recent work, Lerdahl and Jackendoff model a musical work as a trajectory through a 'state space' (involving relative distances between pitches and key signatures). One of the tasks they set themselves is that of figuring out how metric (that is, meter) interpretations are assigned to musical works and what the constraints are on these. We needn't go into the details of these rules, but the overall scheme involves the search for patterns that will function as a grouping mechanism in musical material (i.e. the musical surface structure). These patterns allow expectations to emerge, which in turn allow listeners to make predictions about what is to come.

Another popular theory of music, intended to satisfy this condition of wide explanatory scope, is David Lewin's 'Transformational Theory' (as expounded in his *Generalized Musical Intervals and Transformations*, 2007). As above, according to this approach the 'target system' for the theory is the musical experience of an observer/agent. What is modelled are relational aspects of this experience involving not just relations between pitches (standard intervals), but relationships (generalised intervals) between a host of musical objects.[20] The transformational aspect is supposed to refer to the active procedures carried out by an observer (now conceived as agent). This fairly metaphysical scheme sits atop a very formal bed of representational machinery: musical objects are represented by sets, and the intervals and transformations are members of groups. Hence, group theory provides the mathematical backbone. A musical work is analysed in two ways: passively or actively. The passive approach merely measures the various kinds of intervals, while the active approach uses the notion of a motion shifting one musical configuration to another (providing the dynamics). Again, there is a sense of traversing a space of musical possibilities, to which we will return in a moment.

Lewin's GIS is an example of a musical structure. This gives us something more or less precise that can figure in a scientific theory of music, but it leaves much unsaid. What is the representation relationship between this musical possibility-space and reality? Unlike the situation with physical theories, a 'subjective vs. objective' ambiguity infects even this clean system. Is it a model of the world or the mind, or something else? There is also a curious sense in which it is possible that there is so much theory-ladenness in this procedure that the theory generates the phenomena it is supposed to be describing (a kind of 'performative biolooping') – similar complaints can be made of Schenker's and Lerdahl and Jackendoff's approaches. However, this basic approach, involving group theoretic and geometrical (pitch space) ideas, has been extended by Dmitri Tymoczko in a way that eliminates some of this performativity.

Tymoczko's work is closely related to work on the psychology of pitch relations, as carried out by Carol Krummhansl, among others. The space of musical possibilities is a kind of sensory space, since it is heavily dependent on pitch perception. Such spaces encode implicit (or explicit, for trained musicians) knowledge about musical structure. Matching the persistent ambiguity over whether music theory involves objective or subjective structure, the structure of such musical space/s has been analysed in two ways: 'objectively', via psychology and neuroscience, and 'subjectively', via music theoretical techniques (often based on a listener's introspective experience of a piece of music). Central to both approaches is the notion of tonality.

There is a well-known set of experiments, known as 'probe tone tests', designed to examine a listener's representation of tonality (Krumhansl, 1990). The experiment involves priming the subject with some musical example (from the tonal repertoire), providing a background context and then supplying the subject with a probe (a single tone taken from the 13 chromatic tones of the octave, with the octave tone included here). The subject must then make a judgement about the 'goodness of fit' (relative to a seven-point scale) with the earlier musical passage. The results suggest a fairly robust tonal geometry that matches music theoretical analysis (as represented in, e.g., the circle of fifths). As Krumhansl notes, 'tones acquire meaning through their relationships to other tones' (1990, p. 370). In other words, musical context affects the perception and representation of pitch – so that multiple mental representations map to one and the same physical source. Tonic tones are perceived as 'closest', then diatonic, then non-tonic – this generates a tonal hierarchy. It is precisely the interplay of tonal stability and instability that generates musical tension (produced by motions away from the tonal center) and release (produced by motion back to tonal center), just as is laid out in Lerdahl and Jackendoff's theory.

Krumhansl's (ibid., pp. 112–19) space provides a geometrical model of the perceptual relations among the various keys. An important phenomenological factor that affects the topology of the representational space is octave equivalence ('the sense of identity between tones separated by octaves', ibid., p. 113) – this involves a doubling of frequencies: $f \sim 2f$. Such octave equivalent tones are represented by one and the same point (so that for any pitch p, any pitches that are integer multiples of 12 away [where each semitone amounts to one unit of distance] are considered 'the same').[21]

In Krumhansl's 'probe tone test' for a given C major context, the tonic sits at the vertex, with distance away from the vertex representing goodness of fit: the further away, the poorer the perceived fit with the context (Krumhansl, 1990).

Dmitri Tymoczko (2011) has recently built these basic ideas up into a detailed geometrical framework for doing musicology, examining the 'shapes' of such musical spaces (and more complex counterparts with extended symmetries) and their music theoretic (and musicological, historical analytic) relevance. As above, rather than thinking in terms of basic pitch space (simply: a space whose points represent different pitches ordered in the traditional linear way, such that a musical work traces a path through it), following Krumhansl, Tymoczko identifies *the same* pitches (e.g., middle C, C above, C below and all other Cs), producing the 'pitch-class' spaces rooted in octave-equivalence. Tymoczko recognizes, however, that now the corresponding space is mathematically an example of an 'orbifold' (an orbit-manifold, where the manifold has been 'quotiented' by octave equivalence, a symmetry, thus identifying certain points producing a 'non-simply connected manifold').

In an important extension, Tymoczko generalises this idea to *all* intervals and chords. For example, one could move from middle C to E flat by going up or down (and then jump any number of octaves up or down, so long as an E flat is reached) to get the same interval (tonally/harmonically speaking). Hence, these motions are also identified as musically the same and so quotiented out, reducing the size of the musical possibility space some more, thus producing a far more complex orbifold. As mentioned, one can also apply this to *chords* of any type, with each voice of the chord generating another dimension (axis) of the representational space. The same chord will simply be playable in many different ways, and these are to be identified too, again producing an orbifold when the redundant elements are eliminated through quotienting. There are five such musical transformations for quotienting out redundant structure, which generate the various equivalence classes of 'musical entity' (e.g., chords, chord types, chord progressions, pitch classes, etc.). These are labeled OPTIC symmetries: **O**ctave shifts, **P**ermutation (reordering), **T**ransposition (the relation between pitches sharing the same succession of intervals, regardless of their starting note), **I**nversion (turning a sequence 'upside-down') and **C**ardinality equivalence (ignoring repetitions, note duplication). These are embodied in the structure of the representation space in much the same way that the laws of Newtonian physics are embodied in phase space, with constraints (and symmetries) altering the shape of the space and the kinds of motion that are possible within it (in this physics-based case, with motions representing histories of particles).

Musical compositions can have their tonal structures modelled by such orbifolds, and we can judge the success or failure of compositions relative to them (and the trajectories through them *qua* state spaces). The geometry and topology of the same encodes psychological and physiological constraints (that manifest themselves as the symmetries that reduce the space of possibilities, much as laws of physics generate a dynamical possibility space from a kinematical one). With this model (embodying theory) we have a space of musical possibilities and therefore a means of testing which musical examples will 'work'. Though musicologists have argued that this geometrical approach kills the 'phenomenological complexity of musical experience' (e.g. R. Hasegawa's 'New Approaches to Tonal Theory', 2012), we have seen how at least aspects of this phenomenological experience filter in to constrain the shape of the possibility space.[22]

Meter, like tonality, is clearly a central part of the structural representation of a piece of (Western) music, but in terms of the neural mechanism that supports it, it is really a form of entrainment (that is, the synchronization of internal biological features with external aspects of the environment). This inner/outer link is what causes bodily movements to become coordinated with music. As Justin London puts it, '[M]eter is not fundamentally musical in its origin... [Rather, it] provides a way of capturing the changing aspects of our musical environment as patterns of temporal invariance' (2004, p. 4). Indeed, there are good reasons for believing that meter (and so rhythm) in music is a spandrel of an evolved capacity to model and predict the timing of events. The brain seeks regularities and patterns (invariances) in its environment to reduce cognitive load (energy minimization). However, as a biologically evolved mechanism (a feature of us rather than 'the music itself', whatever that might mean), it is subject to the limitations of biology that brings with it certain (average) temporal limits on processing and action.

London (ibid., p. 190) notes that there exists a 'temporal window' of between 0.1 – 5.0 seconds (the 'specious present') that determines possible metrical cycles.[23] In other words, just as listener experience and capacities inform the modelling and theory of tonality, so they infect the modelling of musical meter: the measure (the units or subharmonics of the beat) have to be within the limits of the present moment, lest it get bundled in memory. So the cycles in the meter must be compressed within the unit, especially for dance responses to occur – London also finds the threshold for perception of *tactus* (the level of beats that are *conducted* and at which we naturally coordinate body movement, such as foot-tapping) at 200 ms to 2,000 ms (ibid., pp. 31–3).[24]

Very recent work[25] shows that meter emerges from the entrainment of neuronal populations resonating to the frequency of the beat, and at the subharmonics corresponding to the metric interpretation of this beat (in ecologically or 'real' valid music). Neural entrainment to beat and meter can be captured directly in the human EEG as a periodic response (manifesting a signal in the EEG) entrained at the frequency of the beat and meter, respectively.[26] For example, if the beat frequency is $f = 2.4$Hz, then binary/march meter (duple time) will be $f/2 = 1.2$Hz and ternary/waltz meter would be $f/3 = 0.8$Hz. On a single, given audible beat frequency (the auditory stimulus), it was found that subjects could induce a 'voluntary metric interpretation' of this beat (with no accents) as binary or ternary (with the appropriate accents for that meter), which in turn induced a further periodic signal in the EEG at the appropriate subharmonic of beat frequency. The idea that musical meter amounts to this kind of entrainment, with the human-side signal-processing (i.e. adding accents that don't exist 'in reality') is known as 'the resonance theory' (for beat- and meter-processing). Like tonality perception, which involves various cognitive inputs onto a musical signal (which themselves contribute heavily to the music that is composed: the object of analysis), meter poses problems in making sense of what the theories and models are about, and as such what music theory is about. This involves an account of what 'musical reality' is. We turn to this in the final section.

4 Musical reality

> [M]usic theory and music itself is the creation of human beings. What does 'reality' mean in such a context?
>
> (Guerino Mazzola[27])

As we have seen, some but not all of the objective structure of music will have subjective correlates. Some but not all of the subjective structure of music will have objective correlates. This lack of isomorphism between the inner and the outer of music is responsible for many of the difficulties and controversies surrounding what a theory of music should be. It ought really to be involved in philosophers' debates about the ontology of music, but isn't... In this final section, we consider what kind of interpretive stance can fit music theories in which the objective and subjective components are all mixed up in this way.[28]

This issue in fact relates superficially to a topic that has been discussed in some detail by philosophers: the extent to which listening to music is conceptual or given (see DeBellis, op. cit. for a good discussion). To pose this issue, consider the following question. Suppose we could give a perfect sonic replication of Bach's music from his own period: would we hear it as they did? The idea is to invoke a problem similar to that raised by the notion of 'incommensurability' in the philosophy of science, according to which users of the theory of one paradigm cannot understand (see the world in the same way) as users of the theory of a different paradigm. In the musical case, the context is the idea of 'historical performance' and the notion that by playing on the instruments of the past (an approximation of sonic replication) one can get closer to a historical listener's *experience*. It is clear from the above discussion that listening to music is by no means simple, and it involves all kinds of cognitive equipment and symmetries. But if these faculties were shared with historical listeners, then presumably they would hear the music in the same way, at this low level, at least. As far as the higher (phenomenological) level, I tend to side with Roger Scruton that historical performance 'cocoon[s] the past in a wad of phoney scholarship, to elevate musicology over music, and to confine Bach and his contemporaries to an acoustic time-warp' (op. cit., p. 448).[29]

One way to frame this with the kind of issue I have in mind, distinct though related to this other debate on conceptualisation of music, is this: do we *hear* meter and tonality? Does it come into our heads from outside of our heads? My preferred response (though not the details of interpretation) has already been revealed (we hear neither in this sense), but it is not standard amongst music theorists. Justin London distinguishes between two camps (with respect to the ontological status of musical meter), the 'structuralist' and the 'phenomenologist':

> The structuralist regards music as existing 'out there,' apart from the listener [what I called a 'worlder' – DR], and thus treats our listening and cognition experiences as our efforts to understand these external sound objects. [...] By contrast, the phenomenologist regards musical structure(s) as the product of the interaction between a sound object and our cognitive faculties [what I called a 'minder' – DR]; she disdains the notion that music qua *music* is only an external sound object, separate from the listener. ... While meter is *not* part of the sound object, it nonetheless may still be regarded as 'part of the music.'
>
> Meter is neither a parameter like pitch or timbre, nor is it part of a nested measuring of durational patterns and/or periodicities. It is something that is heard and felt. And this is of course why the physicist has so much trouble with meter, for physics is not phenomenology. The physicist's job is

> to describe the structure of physical objects in the world. Understanding our interaction with those objects is beyond the scope of the physicist's mission– at least if we stay above the quantum level.
>
> ('Loud Rests and Other Strange Metric Phenomena [or, Meter as Heard]', 1993)

I agree with most of this, though would quibble with the choice of labels. However, it is not true that physics has no frame of reference for dealing with such items as meter and tonality (and therefore, by extension, music itself). Recent work, especially in foundations of physics research, has begun to focus on the role of the observer (especially in the context of quantum mechanics and cosmology). Here we might recall the 'self-excited circuit'of John Wheeler, representing the idea that the universe 'bootstraps' itself into existence by making observing itself, thus creating 'phenomena' (see Wheeler 1980, p. 362). This refers to the fact that, in quantum theory, at least, our observations determine the very reality we are studying (by choosing which experimental questions to put to nature), so that we are in effect studying aspects of ourselves when we examine the quantum world. I want to suggest that something similar can be said of musical reality, and a position similar to Wheeler's was developed ('selective subjectivism'[30]), again in the context of physics, much earlier by the astrophysicist Arthur Eddington.

Selective subjectivism is an epistemological theory that explicitly incorporates 'observer-selection effects' (the idea that in some sense our presence as observers influences *what* we observe).[31] It was intended to be employed in fundamental physical theory, but I propose it here as a novel position in musicology. The position is rather nicely characterised by one of Eddington's famous quotes on the subject:

> We have found a strange footprint on the shores of the unknown. We have devised profound theories, one after another, to account for its origins. At last, we have succeeded in reconstructing the creature that made the footprint. And lo! It is our own.
>
> (1921, p. 201)

In music theory we often think that we are probing some external and eternal realm, but what is happening here is that there is an invisible net underlying our theories, the musicological laws (for want of a better expression); so what is really being studied is the constitution of this net, and this is based on us and our cognitive and physiological makeup. The object of study (music) is entangled with the studiers (the observer). This is a selection effect: An observation-selection effect exists when some feature of a thing is correlated with the observer studying the thing (an example of 'anthropic reasoning').

In fact, there is an even more relevant quote that was written specifically to describe selective subjectivism, based on the concept of a net:

> Let us suppose that an ichthyologist is exploring the life of the ocean. He casts a net into the water and brings up a fishy assortment. Surveying his catch, he proceeds in the usual manner of a scientist to systematise what it reveals. He arrives at two generalisations:
>
> (1) No sea-creature is less than two inches long.
> (2) All sea-creatures have gills.
>
> These are both true of his catch, and he assumes tentatively that they will remain true however often he repeats it.
>
> In applying this analogy, the catch stands for the body of knowledge which constitutes physical science, and the net for the sensory and intellectual equipment which we use in obtaining it. The casting of the net corresponds to observation; for knowledge which has not been or could not be obtained by observation is not admitted into physical science.
>
> An onlooker may object that the first generalisation is wrong. 'There are plenty of sea-creatures under two inches long, only your net is not adapted to catch them.' The icthyologist dismisses this objection contemptuously. 'Anything uncatchable by my net is ipso facto outside the scope of icthyological knowledge.' In short, 'what my net can't catch isn't fish'.
>
> (Eddington, 1938, p. 16)

And we can likewise say, by analogy, that what our net can't catch (i.e. what falls outside the scope of our cognitive powers, etc.) is not music. And, moreover, the catch (the musical object of analysis), to a large extent, takes the form it does on account of the constraints imposed by the net. Here, Marvin Minksy is close to what I take to be an important truth: '[M]usic theory is not only about music, but about how people process it' and 'to understand any art, we must look below its surface into the psychological process of its creation and absorption' (1993, p. 328). But Minsky thinks understanding comes from knowing how something is made. Here, however, absorption and creation go hand in hand, as a result of cognitive/neurobiological features of brains/observer: the creation aspect is *informed* by the absorption aspect. These aspects are well-known amongst music theorists, especially those that favour a more psychological approach. As a pair of apposite passages, consider the following:

> [O]ne might argue that Schenkerian structures characterize both composition and perception. In support of this view, one might argue that the purpose of music, after all, is communication. Why would the great composers have bothered to create such elaborate mental structures if they thought that these structures would never be shared by listeners?
>
> (Temperley, 2011: p, 147)

> [T]he various capacities and thresholds that have been studied and quantified in the psychological laboratory are commonsense aspects of everyday

> musical practice. They are intuitively known to composers when they write their music and to performers in their choices of tempo.
>
> (London, 2012, p. 198)

There is some precedent in the older music theory literature too, in the form of Francois-Joseph Fétis, who wrote the following oft-quoted passage:

> The ear perceives the sounds; the feelings find a priori the formulae of their associations, the mind compares their relationships, judges them, and determines the melodic and harmonic conditions of a tonality.
>
> (F-J Fétis, *Trait complet de la theorie et de la pratique de l'harmonie, contenant la doctrine de la science et de l'art,* 1867)

This is an over-used quotation, but I think he had in mind something like the kind of view I present. This passage captures the idea of the mind's 'selective' action on objective structure (pitch and other materials), rather than passively registering that structure: this is essentially selective subjectivism (it clearly has strong links to Kant, and indeed Fétis was explicit about his debt to Kant, claiming that his was heralding a similar 'Copernican revolution' in music – the 'laws of music' are not derived from the world, but imposed upon them by universal faculties of the mind). Christensen (2004) argues that Fétis viewed it as an *a priori* principle, according to which the mind responds to sensory experience, and viewed it as being of a purely metaphysical nature: 'tonality revealed itself before Fétis's eyes as a logical hierarchy of attractive relationships conceived and imposed upon selected pitch materials by the autonomous intellect, not some external object established by nature that we passively apprehend' (ibid., p. 39). I fail to see how this implies that tonality is metaphysical, but in any case, like Eddington, Fétis too was largely rejected as overly speculative and overly *a prioristic.*

However, in this way, Fétis accounts for music as a merging of objective (external) and subjective (interior) forces. Likewise, Eddington's *a priori* scientist does not introspectively analyse the contents of his head, nor blindly observe the external world, but instead observes observers and the necessary conditions demanded by the construction of whatever class of observables or phenomena interest him.

We can refer back to the work on meter to see how the Eddingtonian framework provides a good fit for the ontological ambiguities that plague music theory. What is curious about meter, you will recall, is that it is part of the essential structure of a piece of music, and yet we do not directly hear meter. There is some strange sense in which the mind is active in creating the piece of music, in contributing an essential part. And yet, in being part of musical structure, it is taken up in music theories: an aspect that comes from the active participation of humans is projected onto the theory itself. We end up modelling aspects of our own constitution (Eddington's net). As London nicely puts it, 'Under this framework meter is a listener-generated construct that is intertwined with the musical surface. Meter is not "part of the music" in the same way that pitch, timbre and duration are. This commitment may be more troubling for some theorists than others...' (London, 1993). For Eddington, 'whatever is accounted for epistemologically [such as the understanding of tonality and meter obtained by observing observers – DR] is ipso facto subjective; it is demolished as part of the objective world' (loc. cit., p. 59) – prime material for selective subjectivist interpretation. Let us finish by paraphrasing another famous quote of Eddington's:

> The ~~physicist~~ [musicologist] might be likened to a scientific Procrustes, whose anthropological studies of the stature of travellers reveal the dimensions of the bed in which he has compelled them to sleep.
>
> (1936, p. 326)

Though physicists did not like this imposition of the mind on reality, it seems much less controversial (and, indeed, mandatory) to allow this in the case of musical reality.

5 Conclusion

Regardless of whether or not selective subjectivism is to be taken seriously as a philosophical stance in the case of music and the interpretation of music theory, it can readily be seen that music theory (along with the peculiarities of certain structural elements of music) contains several areas of interest for philosophers of science. Indeed, the awkwardness of music as revealed in tonality and meter studies (neither fully subjective nor fully objective) might provide new data for those (naturalistic) philosophers interested in the ontology of music and who wish their ontology to be continuous with the best results in the theory of music.

A remaining problem, also one that philosophers of science might contribute to (and that is important for music theorists), is the question of what a theory is supposed to be modelling. There are two levels to this problem: first, is it the objective structure in the sound signal itself or is it the experiences of observers (listeners)? If the latter, then we have further complications, since it is not clear which listeners we have in mind: untrained or trained? How well trained? An ideal, perfectly trained one? In this brief review, I have been most concerned not with describing and analysing the details of music theories (which is an enormous subject), but in highlighting those tensions and issues that I find might be of most interest to philosophers of science.

Notes

1 These temporal aspects have been studied quite deeply in the context of a phenomenological (especially Husserlian) study of music – with the concepts of 'retentions' and 'protensions' (essentially the projection of past and future-expected events into present

consciousness) playing a central role in explaining how a work of music (with its long, melodic spans and temporally extended structure) can be experienced in a present moment. See, especially, Izchak Miller (1984). For a general bibliography of works dealing with the relationship of time and music, see Jonathan Kramer (1985).

2 Note that often, in philosophy of science, when we think of 'theories' we think of the laws that form its basis (for example, Newton's Laws of Motion). Such laws are expressed through symmetries (for example, Galilean symmetries in the case of Newtonian physics), so I'll be ultimately thinking of the search for the laws of music as the search for invariances in music. Theories will then be characterised with reference to these laws.

3 Matthew Brown, a musicologist, has tackled (in his *Explaining Tonality: Schenkerian Theory and Beyond*, University of Rochester Press, 2005) the problem of explaining tonality (among other methodological issues) using tools from philosophy of science, but this study (and his other related work) appears to be an isolated exception. Two other books skirting philosophy of science issues are Mark DeBellis' *Music and Conceptualization* (Cambridge University Press, 2008) and Charles Nussbaum's *The Musical Representation: Meaning, Ontology, and Emotion* (MIT Press, 2007). Roger Scruton's *The Aesthetics of Music* (Oxford University Press, 1997) also considers, albeit briefly, music analysis from a philosophical standpoint.

4 We might also note that there have been attempts to subsume music (and music theory) under other sciences (especially biology and neuroscience) as a natural phenomenon, much as is the case with economics, theology and other human behaviours and behavioural sciences. For example, the 'Generative Theory of Tonal Music' of Lerdahl and Jackendoff (1983) explicitly treats music theory as a branch of theoretical psychology.

5 Likewise Moritz Hauptmann, who in his *The Nature of Harmony and Meter* (1888, p. xl), writes that: 'That which is musically inadmissible is not so because it is against a rule determined by musicians but because it is against natural law given to musicians from mankind, because it is logically untrue and of inward contradiction'.

6 As mentioned above, there are some overlaps with the scientific study of time here, in which it is ambiguous as to whether such features as 'flow' are 'in the world' or 'in the head'. Indeed, Zuckerkandl (1969) argues that music enables us to experience time as a concrete entity and, in some sense, highlights the reality of time. Comparing music to the 'visual arts' (spatial), Marvin Minsky wonders, given that music seems to have some stability to it when we experience it, 'How [this can] be, when there is so little of it present at each moment?' (1981, p. 338). That is, as with more general experience (which has unity), music brings the problem of integrating experience across time into stark relief.

7 I shall refer to these two camps as 'minders' and 'worlders' respectively. The Eddingtonian position brings the two together in a way that seems entirely appropriate to the curious nature of music.

8 Music theory is more closely related to 'music analysis' in this respect, though the latter focuses more on individual pieces of music, rather than *principles of music* in general. Indeed, music analysis can be used to test theory, not offering predictions about other works but perhaps functioning as a confirmation of some music theory. As Ian Bent puts it, music theory consists of 'that part of the study of music which takes as its starting point the music itself, rather than external factors' (1980, p. 341). Of course, this also faces the ambiguity previously mentioned, concerning whether 'music itself' is something objective or subjective (or something else). There is a sense in which this distinction between music theory/analysis and musicology can be mapped to 'internalist' and 'externalist' approaches in the history and philosophy of science.

9 Roger Scruton argues that the (cross-temporal) harmonic, rhythmic or melodic organization of sound is what distinguishes a musical experience from a merely sonic one (2007). This grouping of sounds into units and patterns is an essential part of the musical experience.

10 Relative pitch most likely has a basis in 'intonation' perception in ordinary speech (e.g. a rise in pitch indicates a question, at least amongst non-Australian English speakers).

Those that are without 'music sense' (i.e. sufferers of amusia) also have a marked inability to process the intonation of speech: see P. Podlipniak's 'The Ability of Tonal Recognition as One of Human-Specific Adaptations' (2015: §4).

11 The octave is divided into 12 semitones (in most Western music). This octave division is the foundation for musical scales: we divide to preserve consonance, which, of course, has a psychological element: consonance refers to what is pleasing to *us*! Harmony is essentially the study of the mathematical relations forming the structure of music that we find pleasing. Notes possess certain basic frequencies (specified in cycles per second, or hz), so that, for example, A = 440hz. The ratio between frequencies is essential to harmony; we achieve consonance when this frequency ratio is a ratio of small integers: 1:1 = unison; 2:1 = octave (880hz:440hz); 3:1 = perfect 5th and so on.

12 In more detail, 'tonal' refers to a class of music in which the organisation of pitches singles out some special pitch (the tonic) which serves to pick out relative consonance and dissonance with respect to other tones. By contrast, atonal music involves the 'method of composing with twelve tones which are related only with one another' (A. Schoenberg, 1984, p. 218), so that the twelve pitches of the octave (unrealized compositionally) are regarded as equal, and no one note or tonality is given its classical harmonic significance. Schoenberg, who first developed this idea, regarded it as the musical equivalent of Einstein's relativity principles, much like the principle of background independence: no tonal background, only relative tones.

13 This is a well-known phenomenon, though the entanglement of cause and effect has not been adequately discussed. Here's Justin London: 'At some times a sense of accent flows from the musical surface to the emerging meter, and at other times from the meter to the unfolding musical surface (1993, p. 10). Or take Victor Zuckerkandl: 'It is not a differentiation of accents which produces meter, it is meter which produces a differentiation of accents' (1956, p. 169).

14 That is, in terms of structures ⟨M,Oi⟩ consisting of a background manifold (a set of spacetime points with a topology and possibly a metric defined on them) and a set of geometric objects representing the kinds of matter one wishes to represent (e.g., particles, fields or strings).

15 As Brown *et al.* put it: 'If Schenkerian theory [or replace with any other theory – DR] turns out to be powerful and fruitful because, like creationism, it is a universal technique that yields analytical results for any sequence of pitches whatsoever, then it is hardly worth regarding as a theory at all, much less a theory of tonality. Such a theory would be complete, but wildly unsound. A good theory is one that is not universally extendible to all music' (1997, p. 158).

16 Though it might be a stretch, there is some similarity between this situation and the way in which quantum states are conceived in the QBist approach to quantum mechanics, in which both the perceiving subject and the perceived object are considered central. See N. D. Mermin's 'QBism puts the Scientist Back in Science'. (2014: pp. 421–3).

17 The authors are well aware of the idealization involved in their experienced listener (see, e.g., ibid., p. 3), and they even introduce the concept of 'a perfect listener' (a '*homo musicus*' which puts one in mind of the *homo economicus* from neo-classical economic theory). However, they take the composer to have in mind one who knows what they are doing...

18 Milton Babbitt has written that 'Schenker has contributed..., a body of analytical procedures which reflect the perception of a musical work as a dynamic totality, not as a succession of moments or a juxtaposition of "formal" areas related or contrasted merely by the fact of thematic or harmonic similarity or dissimilarity!' (1952, p. 22). Schenkerian theory is really the default in musicology: it's a recursive theory in which musical structure is analysed in terms of successive pitches (voices) elaborating on previous ones. Levels of structure in a piece with a main 'urlinie' structure featuring stepwise descent, from mi, re, do usually (but sometimes this can be more complex: sol, fa, mi, re, do, etc...). Lots of Husserlian ideas can be found in Schenker's theory, and it is based

in a key way on 'retention' and 'protention' in analysing musical works – this is why philosophers of time are interested in this work, as mentioned earlier.

19 For example, the movement away from the tonal centre of a piece can generate a sense of tension, which can be resolved by motion back towards the tonal centre, achieving closure when the tonic is reached again.

20 We can precisify this a bit via Lewin's generalised interval system [GIS], so that there is a basic set of musical/sonic-objects: pitches, durations, intensities (the raw materials of music). Of course, he also postulates a group of generalised intervals, linking these basic musical elements. As mentioned, Lewinian intervals are far more than distances in pitch or frequency ratios; rather, they model, among other things, transpositions between pitches, pitch classes, chords and series; serial transformations; and rhythmic relationships. There is then a mapping [the 'interval function'] that spits out an interval for each ordered pair of musical objects.

21 Mathematically, this reduction of the space of musical possibilities (by identifying 'musically identical' tones) corresponds to shifting to the reduced space $\mathbb{R}/12\mathbf{Z}$ (the reals with the set of multiples of 12 'quotiented out'), and is what music theorists call 'pitch class space' (cf. Tymoczko, 'The Geometry of Musical Chords', 2006, p. 72). *Sameness* here is couched in terms of 'musical structure' (though in such a way as to be related to the listener's experience, such as preserving the character of a chord in a transposition, for example). In this sense, octave equivalence is a symmetry of the model.

22 There is neuroscientific data that maps the kinds of results obtained in music theory, by Lerdahl and Jackendoff, Krumhansl, Tymoczko and others, onto specific brain areas. Janata, *et al.* (2002) found, in fMRI experiments 'an area in rostromedial prefrontal cortex that tracks activation in tonal space' mapping onto the 'formal geometric structure that determines distance relationships within a harmonic or tonal space' in the specific context of Western tonal music, where '[d]ifferent voxels [3D pixels] in this area exhibited selectivity for different keys' (ibid., p. 2167). This evidence supports the idea that our neural representations of music are relational: on a specific occasion some family of neurons will fire, say for the key of A minor, while on another occasion that family might fire for the key of C minor. But the fact remains that the relational structure between the keys is preserved on each occasion, despite the fact that keys are not absolutely localised to unique assemblies of neurons, and that the symmetries of the model are mapping to symmetries in reality. As the authors put it: 'what changed between [fMRI] sessions was not the tonality-tracking behaviour of these brain areas but rather the region of tonal space (keys) to which they were sensitive. This type of relative representation provides a mechanism by which pieces of music can be transposed from key to key, yet maintain their internal pitch relationships and tonal coherence' (ibid., p. 2169). Again, this is closely related to the way in which symmetries in physics lead to relative representations of physical quantities. (For more on the philosophical significance of these observations, see Diana Raffman's 'Music, Philosophy, and Cognitive Science', 2011).

23 In his words: '[T]he shortest interval that we can hear or perform as an element of rhythmic figure, is about 100 milliseconds (ms) [and] the upper limit is around 5 to 6 seconds, a limit set by our capacities to hierarchically integrate successive events into a stable pattern'.

24 Some musics have repeating patterns that exceed this threshold, e.g. the *tal* cycles of Indian raga. As the threshold is exceeded, other mechanisms come into play (see Martin Clayton, op. cit. for more on this complication).

25 E.g. A. Tierney and N. Kraus, 'Tagging the Neural Entrainment to the Rhythmic Structure of Music', 2015.

26 As with tonality, a number of electrophysiological experiments reveal that humans have a marked preference for integer ratios in meter perception (see e.g., Pablos Martin *et al.*, 'Perceptual Biases for Rhythm: The Mismatch Negativity Latency Indexes the Privileged Status of Binary vs. Non-Binary Interval Ratios', 2007).

27 http://www.encyclospace.org/special/answer_to_tymoczko.pdf, p. 4.

28 Note also that to a large extent how we interpret basic musical structure will determine the kinds of theories we generate, so that philosophical work along these lines can have a direct bearing on music theory itself.

29 But there is, of course, something to the strongly conceptual content of any experience. As Goodman and Elgin note: 'A particular auditory event might be heard as a noise, as a piercing noise, as the sound of a trumpet, as a B flat, as the first note of a fanfare or in any of indefinitely many other ways. To characterise what is heard as the sound of a trumpet or as the first note of a fanfare requires a good deal of background knowledge. But every characterisation relies on background knowledge of one sort or another. Even to recognise something as a sound requires knowing how to differentiate sounds from other sources of sensory stimulation, and how to segment auditory input into separate events. Sensation is sometimes supposed to be primarily given' (1988, pp. 9–10). Here I am more interested in the mind's projections onto incoming signals that operate at a lower level.

30 In brief, in the words of its inventor: 'Selective subjectivism, which is the modern scientific philosophy, has little affinity with Berkeleian subjectivism, which, if I understand it correctly, denies all objectivity to the external world. In our view the physical universe is neither wholly subjective nor wholly objective—nor a simple mixture of subjective and objective entities or attributes' (1938, p. 27).

31 Here is a quick example of a selection effect known as the 'fine-tuning problem': in cosmology, our thinking about the nature of the universe must take account of the fact that we are in a relatively special epoch of the universe's history in which complex organisms are possible, and slight differences (in, e.g., the values of the constants of nature) would forbid our existence. If such observers can exist only in universes in which the relevant parameters (gravitational strength, etc.) take on the observed 'fine-tuned values', then an observation selection effect can be used to explain why we observe a fine-tuned universe: if they didn't have these values, we wouldn't be there to observe them. Ditto with music: Why does music have the features we observe? Because our presence as observers with a certain constitution (with certain temporal thresholds on our conscious awareness, the presence of symmetries and so on) necessitates certain features (the 'discovered' features) to be present in the music.

References

Babbitt, M. (1952). 'Structural Hearing', in S. Peles, *et al.* (eds.), *The Collected Essays of Milton Babbitt*. Princeton, MA: Princeton University Press.

Bent, I. (1980). 'Analysis', in S. Sadie (ed.) *The New Grove Dictionary of Music and Musicians*. London: Macmillan.

Brown, M. (2005). *Explaining Tonality: Schenkerian Theory and Beyond*. Rochester, NY: University of Rochester Press.

Brown, M., D. Dempster and D. Headlam. (1997). 'The #IV(♭V) Hypothesis: Testing the Limits of Schenker's Theory of Tonality'. *Music Theory Spectrum*, 19 (2), pp. 155–83.

Burnham, S. (1992). 'Method and Motivation in Hugo Riemann's History of Harmonic Theory'. *Music Theory Spectrum*, 14 (1), pp. 1–14.

Christensen, T. (2004). *Rameau and Musical Thought in the Enlightenment*. Cambridge: Cambridge University Press.

Clarke, E. (1987). 'Levels of Structure in the Organization of Musical Time'. *Contemporary Music Review*, 2 (1), pp. 211–38.

Clayton, M. (2000). *Time in Indian Music*. Oxford: Oxford University Press.

DeBellis, M. (2008). *Music and Conceptualization*. Cambridge: Cambridge University Press.

Eddington, A. (1921). *Space, Time, and Gravitation.* Cambridge: Cambridge University Press.

Eddington, A. (1936). *Relativity Theory of Protons and Electrons.* Cambridge: Cambridge University Press.

Fétis, F. -J. (1867). *Trait complet de la theorie et de la pratique de l'harmonie, contenant la doctrine de la science et de l'art.* Paris: Maurice Schlesinger.

Goodman, N. and C. Elgin. (1988). *Reconceptions in Philosophy and Other Arts and Sciences.* Cambridge, MA: Hackett Publishing.

Harap, L. (1937). 'On the Nature of Musicology'. *The Musical Quarterly,* 23 (1), pp. 18–25.

Hasegawa, R. (2012). 'New Approaches to Tonal Theory'. *Music and Letters,* 93 (4), pp. 574–93.

Hauptmann, M. (1888). *The Nature of Harmony and Meter.* London: Swan Sonnenschein.

Janata, P., Birk, J. L., Van Horn, J. D. Leman, M. Tillmann, B. and Bharucha, J. J. (2002). 'The Cortical Topography of Tonal Structures Underlying Western Music'. *Science,* 298, pp. 2167–70.

Kramer, J. (1985). 'Studies of Time and Music: A Bibliography'. *Music Theory Spectrum, 7 (Time and Rhythm in Music),* pp. 72–106.

Krumhansl, C. L. (1990). *Cognitive Foundations of Musical Pitch.* Oxford: Oxford University Press.

Lerdahl, F. and R. Jackendoff. (1983). *A Generative Theory of Tonal Music.* Cambridge, MA: MIT Press.

Lerdahl, F. and R. Jackendoff. (2001). *Tonal Pitch Space.* Oxford: Oxford University Press.

Lewin, D. (2007). *Generalized Musical Intervals and Transformations.* Oxford: Oxford University Press.

London, J. (1993). 'Loud Rests and Other Strange Metric Phenomena (or, Meter as Heard)'. *Music Theory Online, 0 (2).* Available at: http://www.mtosmt.org/issues/mto.93.0.2/mto.93.0.2.london.art.

London, J. (2012). *Hearing in Time.* Oxford: Oxford University Press.

Martin, P., Pablos Martin, X., Deltenre, P., Hoonhorst, I., Merkessis, El. Rossion, B. and Colin, C. (2007). 'Perceptual Biases for Rhythm: The Mismatch Negativity Latency Indexes the Privileged Status of Binary vs. Non-Binary Interval Ratios'. *Clinical Neurophysiology,* 118, pp. 2709–15.

Mermin, N. D. (2014). 'QBism Puts the Scientist Back in Science'. *Nature,* 507, pp. 421–3.

Miller, I. (1984). *Husserl, Perception, and Temporal Awareness.* Cambridge, MA: MIT Press.

Minsky, M. (1993). 'Music, Mind, and Meaning', in S. M. Schwanauer and D. A. Levitt (eds.) *Machine Models of Music.* Cambridge, MA: MIT Press.

Mirka, D. (2009). *Metric Manipulations in Haydn and Mozart.* Oxford: Oxford University Press.

Nussbaum, C. (2007). *The Musical Representation: Meaning, Ontology, and Emotion.* Cambridge, MA: MIT Press.

Podlipniak, P. (2015). 'The Ability of Tonal Recognition as One of Human-Specific Adaptations', in C. Maeder and M. Reybrouck (eds.) *Music, Analysis, Experience: New Perspectives in Musical Semiotics.* Leuven, Belgium: Leuven University Press.

Raffman, D. (2011). 'Music, Philosophy, and Cognitive Science', in T. Gracyk and A. Kania (eds.) *The Routledge Companion to Philosophy of Music.* London: Routledge.

Rameau, J-P. (1984). *Treatise of Harmony.* Dover: Dover Publications.

Scruton, R. (1997). *The Aesthetics of Music.* Oxford: Oxford University Press.

Scruton, R. (2007). 'Thoughts on Rhythm', in K. Stock (ed.) *Philosophers on Music.* Oxford: Oxford University Press.

Temperley, D. (2011). 'Composition, Perception, and Schenkerian Theory'. *Music Theory Spectrum,* 33 (2), pp. 146–168.
Tierney, A. and N. Kraus. (2015). 'Tagging the Neural Entrainment to the Rhythmic Structure of Music'. *Journal of Cognitive Neuroscience,* 27 (2), pp. 400–8.
Tymoczko, D. (2006). 'The Geometry of Musical Chords'. *Science,* 313, pp. 72–4.
Tymoczko, D. (2011). *The Geometry of Music.* Oxford: Oxford University Press.
Zuckerkandl, V. (1969). *Sound and Symbol* (translated by W. R. Trask). New York: Pantheon Books.

8 Kant on beauty and cognition[1]

Alix Cohen

1 Introduction

Kant often seems to suggest that a cognition – whether an everyday cognition or a scientific cognition – cannot be beautiful. In the *Critique of Judgment* and the *Lectures on Logic* he writes: 'a science which, as such, is supposed to be beautiful, is absurd' (CJ 184 [5:305]). 'The expression [,] beautiful cognition [,] is not fitting at all' (LL 446 [24:708]).[2] These claims are usually understood rather straightforwardly. On the one hand, cognition cannot be beautiful, since by Kant's account it is all about concepts, while beauty is defined by its non-conceptual nature. On the other hand, beauty cannot contribute to cognition, since the former is grounded on subjective feelings while cognition is all about objective knowledge.[3] However, I will argue that Kant's view of the relationship between cognition and beauty is not as straightforward as it may seem, and that both of these claims are in fact false. As I will show, for Kant, cognition can be beautiful, and the feeling of beauty is cognitively valuable. Yet it is not because beauty is a sign of the truthfulness of a theory. Nor is it because the process that gives rise to the feeling of beauty, the free play, furthers scientific progress. Rather, it is because the feeling of beauty stimulates our cognitive powers and thereby enhances our cognitive activity. On this basis, contrary to what is usually thought, cognition can, and in fact should, be beautiful for Kant.

To support this claim, I begin by arguing that if science cannot be beautiful, it is not because it involves the application of concepts. Insofar as, on my account, the same object can be apprehended both cognitively and aesthetically without contradiction, we have no reason to doubt the possibility of an aesthetic dimension of cognition: cognition can be beautiful, although it is not necessarily so. Yet it could be that this dimension is irrelevant to cognition itself. Thus, the question is whether its beauty can be relevant to, and even useful for, our cognitive endeavours. In the second section, I examine the main challenge to the idea of an epistemic contribution of beauty to cognition. According to rationalist interpretations of Kant's account of cognition, feelings are at best irrelevant and at worst a hindrance to our cognitive endeavours, which suggests that for Kant beauty cannot contribute to cognition. The third section discusses an attempt to circumvent this challenge by arguing for the positive role of beauty in cognition on the basis of the epistemic contribution of aesthetic reflection. However, I go on to suggest that the distinction between aesthetic and cognitive reflection rules out this possibility. Insofar as rational model of cognition entails that we are unable to argue for the role of beauty in cognition on the basis of the epistemic function of the feeling of aesthetic pleasure, it seems that Kant's account precludes the idea of an epistemic contribution of beauty all together. However, Section 4 argues that Kant doesn't hold such an account. To support this claim, I turn to his account of the aesthetic dimension of cognition, which encompasses the different kinds of effects cognition has on the faculty of feeling, and show that the beauty of a cognition is epistemically valuable. In this sense, far from portraying cognition as opposed to beauty, we should acknowledge that the aesthetic dimension of cognition has a rightful place in Kant's epistemic framework.

2 The possibility of a beautiful cognition

Famously for Kant, the feeling of aesthetic pleasure that defines judgments of beauty involves what he calls the harmonious free play of imagination and understanding.

> [T]he state of mind in this representation must be that of a feeling of the free play of the powers of representation in a given representation for a cognition in general. … this merely subjective (aesthetic) judging of the object, or of the representation through which the object is given, precedes the pleasure in it, and is the ground of this pleasure in the harmony of the faculties of cognition.
>
> (CJ 102–3 [5:217–8])

This statement has generated ongoing debates in the literature, but for the purpose of this paper, I want to focus on the notion of harmonious free play.

Many commentators work under the assumption that for Kant, harmony and free play are so intrinsically connected that any interplay that is free is harmonious and vice versa. This claim generally leads them to adopt one of two views: either all cognitions are beautiful, or no cognition can be beautiful. According to the former, any object, insofar as it is cognised, is beautiful because it generates a harmonious interplay of our faculties. Since cognising necessarily involves such harmony, nothing can be either aesthetically indifferent or ugly; everything is beautiful.[4] According to the latter, no cognition can be beautiful because it involves determinate concepts and thus inhibits the free play of the faculties, while judgments of beauty are non-conceptual and involve such free play. In other words, there is no beauty in cognition since there is no room for it: cognition is determined by concepts throughout.[5] Both views are problematic in their own right, but I believe they are wrong for the same reason: they share the assumption that harmony and free play are intrinsically connected. In contrast, I will argue that we should distinguish between them because not all cases of harmony are cases of free play and not all cases of free play are cases of harmony. On this basis, I will formulate an account according to which cognition can be but isn't necessarily beautiful.

To support this claim, let's begin by focusing on the contrast between judgments of beauty and cognitive judgments. On Kant's account, cognition and aesthetic experience engage the same faculties, imagination and understanding, 'so far as they agree with each other as is requisite for a *cognition in general*' (CJ 103 [5:218]). These faculties interact in a harmonious fashion, although they do so in different ways. In the case of cognition, their harmonious interplay is restrained by the application of concepts, since for Kant, knowledge consists in the determinate application of concepts to intuitions schematised by the imagination.[6] In contrast, aesthetic experience engages the same faculties but in a reflective rather than a determining fashion. In this case, the harmonious interplay between imagination and understanding is free and no concept is applied.

> The powers of cognition that are set into play by this representation [i.e., an artistic representation] are hereby in a free play, since no determinate concept restricts them to a particular rule of cognition.
>
> (CJ 102 [5:217])

While the contrast between judgments of beauty and judgments of cognition seems straightforward enough, it has been interpreted as entailing that no object of cognition – whether everyday cognition or scientific cognition – can be beautiful because it involves concepts and excludes free play, whereas beauty is non-conceptual and involves free play. Now, there is no doubt that the contrast between beauty and cognition is well-grounded: for Kant, they differ in meaningful and significant ways. What I believe, however, is that this contrast is compatible with the claim that cognition can be beautiful. For none of the characteristics of judgements of beauty entail that cognitions cannot be aesthetically apprehended. This is in fact what Kant alludes to at the very beginning of *Critique of Judgment*: 'even if the given representations were to be rational but related in a judgment solely to the subject (its feeling), then they are to that extent always aesthetic' (CJ 90 [5:204]). What he suggests here, albeit implicitly, is that the same representation can be related to the subject in a number of ways, and which way obtains is defined by the determining ground of the judgment that ensues: it is a cognitive judgment if it is grounded on a concept, and it is an aesthetic judgement if it is grounded on a feeling of disinterested pleasure. Hence, even when we are judging a representation that is fully conceptualised and determined ('rational'), as long as our judgment is based on disinterested pleasure, it is a judgement of taste.

Readers of Kant have resisted accepting this account of the distinction between cognitive and aesthetic judgments because instead of emphasising what distinguishes them from each other (namely, their grounds), they focus on what they have in common (namely, the faculties they engage). In their picture, insofar as the same faculties (i.e., imagination and understanding) are at play in both operations of the mind (i.e., cognitive and aesthetic judgment), their harmonious interplay can only be of one kind 'at any given time'[7] (i.e., it can only be either free or unfree). The nature of cognition thus precludes the aesthetic engagement with its objects.[8] However, not only does Kant's account of cognition give us no reason to accept this picture, it gives us good reason to deny it. For the same faculties can be in unfree harmony in one respect (cognitive, determining judgment) and in free harmony in another (aesthetic, reflective judgment).

To make sense of this claim, let's begin with Kant's well-known example of a flower. While I judge it aesthetically, I am nevertheless aware of the fact that it is a flower of a certain kind, even if I do not pay attention to this fact. As Kant notes,

> Flowers are free natural beauties. Hardly anyone other than the botanist knows what sort of thing a flower is supposed to be; and even the botanist, who recognizes in it the reproductive organ of the plant, pays no attention to this natural end if he judges the flower by means of taste.
>
> (CJ 114 [5:229])

I have successfully applied the concept of, say rose, through determining judgment, to the given of intuition, and in this respect, the imagination and the understanding function in unfree harmony.[9] But this unfree cognitive harmony can obtain while the imagination and understanding are also in free aesthetic harmony, in a judgement of taste. In other words, in the experience of the rose, I experience simultaneously cognitive harmony and aesthetic harmony between imagination and understanding, although they differ insofar as in the former a concept is applied (cognitive unfree harmony) while in the latter no concept is applied (aesthetic free harmony). What these experiences have in common, however, is that they are both pleasurable, although the kind of pleasure they give rise to differs: aesthetic pleasure is disinterested, whereas cognitive pleasure is not. It is the result of the attainment of our cognitive aim.

> This determination [determining judgment] is an end with regard to cognition; and in relation to this it is also always connected with satisfaction (which accompanies the accomplishment of any aim, even a merely problematic one). But then it is merely the approval of the solution that answers a problem, and not a free and indeterminately purposive entertainment of the mental powers with that which we call beautiful, where the understanding is in the service of the imagination and not vice versa.
>
> (CJ 125-6 [5:242])[10]

While determinative cognition gives rise to cognitive pleasure, aesthetic pleasure is sustainably ongoing in its freedom insofar as the reflective process it consists in is itself pleasurable because harmonious. However, contrary to what is often thought, not all reflective processes are either harmonious, pleasurable or free.[11]

To make sense of this claim, let's go back to the example of a rose. Upon encountering a new species of rose, I 'determinately' apply the empirical concept of 'rose', while 'reflectively' looking for a concept distinct from that of 'rose', which I do have, that would better fit the object. In determining judgment, the interplay between imagination and understanding consists in the application of the concept of rose, through judgment, to the given of intuition. Cognitive faculties are in harmony, but an unfree one – what Kant sometimes calls the 'lawful agreement' between the powers of cognition (CJ 175 [5:295]). In reflective cognitive judgment, however, the interplay between imagination and understanding is not unfree in the same way, since no concept is applied and no cognition achieved. But it is not free either, since in this case reflective judgment nevertheless operates 'in relation to a concept thereby made possible' (CJ 15 [20:210]). Cognitive reflection is constrained by the fact that it is conceptually steered: it aims at the determinate application of a concept although it fails to achieve it (or has not achieved it yet). We could say that it is reflectively, objectively purposive, in contrast to both the determinate objective purposiveness of cognition and the reflective subjective purposiveness of aesthetic reflection, which is 'a free and indeterminately purposive entertainment of the mental powers' (CJ 126 [5:242]). Thus cognitive reflection is not harmonious until it stabilises itself in its determinative form: imagination and understanding are in disharmony – or at least not in harmony – and no pleasure arises from their reflective interplay. On this basis, insofar as the cognitive and the aesthetic use of reflective judgment consist of different mental processes, we need to distinguish between them: the former is neither free nor harmonious and, because not harmonious, not pleasurable, while the latter is free and harmonious, and because harmonious, pleasurable.[12]

However, one may object that judgments of beauty and of cognition shouldn't be so different that the aesthetic free play loses its connection to cognition.[13] Otherwise, the universal validity of judgments of taste cannot be accounted for.[14] It is true that on my reading, cognition and beauty do consist in different mental processes so that the latter is not required for the possibility of the former. But recall that for Kant, what 'is requisite for a *cognition in general*' is that 'they [imagination and understanding] agree with each other (*zusammen stimmen*)' (CJ 103 [5:218]) – which I interpret as the requirement that beauty and cognition both engage the same faculties *in a harmonious fashion.*[15] That they harmonise with each other in different ways (freely for beauty and unfreely for cognition) is irrelevant since on my account, it is the harmony that is the ground of the universal validity of judgments of taste. It still remains the case that aesthetic pleasure arises from our fundamental shared cognitive powers functioning harmoniously 'rather than from merely idiosyncratic associations' as Guyer puts it (Guyer (2006): 315). While it doesn't guarantee that every creature endowed with these capacities will feel aesthetic pleasure, it does secure the claim that they have the capacity for it, and thus that 'the pleasure … can rightly be expected of everyone' (CJ 170 [5:290]).

As a result, there is no reason to believe that the harmony required by cognition is necessarily of the same kind as the harmony entailed by the

Table 8.1 Different kinds of judgements

Interplay of the faculties	*Free (conceptually undetermined; subjective)*	*Unfree (conceptually determined; objective)*
Harmonious (accompanied by feeling of pleasure)	Beauty	Determining cognition
Disharmonious (accompanied by feeling of displeasure)	Ugliness	Reflective cognition

experience of the beautiful, and thus that all cognitions are beautiful. Nor is there reason to believe that the unfree play required by cognition rules out the free play required by the experience of the beautiful, such that cognition and the experience of beauty are incompatible. On the contrary, they involve different but compatible and potentially concurrent mental processes or functions – summarised in Table 8.1. On this basis, we are left with no reason to doubt the possibility of an aesthetic dimension of cognition: cognition can be beautiful, although it is not necessarily so.[16] Of course, it could be that this dimension is actually irrelevant to cognition itself. Thus the next question to address is whether its beauty can be relevant to, and even useful for, our cognitive endeavours. The next section will begin with the biggest challenge to the idea of an epistemic contribution of beauty to cognition; namely, we have good reasons for thinking that for Kant, feelings are at best irrelevant and at worst a hindrance to our cognitive endeavours.

3 Kant's supposed rationalist model of cognition

Although little attention has been paid in literature to the question of the role of feelings in Kant's account of cognition, it is usually assumed that for Kant, they do not play a role in the acquisition of knowledge, or that if they do play a role in it, they can only be a hindrance rather than a help. If this assumption is correct, it would seem to entail that feelings of beauty, *qua* feelings, are necessarily irrelevant to cognition if not hindrances to it.

The presumption that Kant holds a rationalist conception of cognition is based on the fact that he defines feelings in terms that seem at odds with the very nature of cognition: they are subjective and contingent affective states while cognition consists in objective and necessary judgments. The few commentators who do mention this issue simply conclude that for Kant, emotions distort cognition since they are illnesses of the mind.[17] On this basis, Kant's view of the relationship between feelings and cognition has been interpreted along the lines of what McAllister has called the 'rational model of science' – a model according to which feelings are irrelevant to our epistemic inquiries.[18] The acquisition of knowledge is a theoretical enterprise that, as such, only necessitates the intervention of cognitive faculties (i.e., theoretical reason, the understanding, sensibility, the imagination and judgment). As Kant himself writes, feelings 'contribute nothing to the play of our representations as powers of cognition' (LL 270 [24:811]).[19] They do not have any cognitive import, since they yield merely subjective certainty: 'frequently we take something to be certain merely because it pleases us [...] This certainty or uncertainty is not objective, however, but instead subjective' (LL 157 [24:198]). Therefore, feelings do not have any epistemic relevance. If they do intervene in our epistemic pursuits, they have 'dangerous consequences' for cognition (LL 129 [24:163]). On my reading these consequences are of two types, what I would like to call 'obstruction' and 'intrusion', and it is important to distinguish between them.

The first type of dangerous consequence is obstruction. Feelings impede upon cognitive faculties and processes, making them less efficient and reliable. They hinder cognition by preventing our faculties from functioning properly: 'Everything that stimulates and excites us serves to disadvantage our power of judgment' (LL 44 [24:60]). They distort our cognitive processes and the acquisition of knowledge: 'stimulation and excitement, most of all, can spoil the logical perfection in our cognitions and judgments' (LL 547-8 [9:37]).[20] The second type of dangerous consequence is intrusion. Rather than merely impeding cognition, feelings intrude in it by introducing a subjective dimension into what should be wholly objective.[21] They prompt us to adopt beliefs on subjective rather than objective grounds, for instance because they suit our taste: 'No aesthetic proof can be a demonstration, then, for an aesthetic probation arises merely out of the agreement of cognitions with our feeling and our taste [;] thus it is nothing but persuation'(LL 186 [24:234]).[22]

When feelings intrude upon cognition, we introduce a subjective dimension that does not belong to the realm of knowledge, thereby leading to cognitive bias, distortion, partiality, etc.[23] Feelings are thus 'foreign powers' that bring non-epistemic concerns to bear onto epistemic ones: 'when foreign powers mingle with the correct laws of the understanding, a mixed effect arises, and error arises from the conflict of [this with] our judgments based on the laws of the understanding and of reason' (LL 79 [24:102]).

Feelings should neither obstruct nor intrude upon cognition, which seems to entail that Kant holds a rationalist account of cognition according to which all feelings, including aesthetic ones, are at best irrelevant and at worst a hindrance to our cognitive endeavours. However, the next section, rather than concluding that beauty cannot contribute to cognition, examines an attempt to circumvent this challenge by moving away from aesthetic feelings and focusing instead on aesthetic reflection.

4 The epistemic function of aesthetic reflection

A number of commentators have suggested that cognitive and aesthetic judgments are mutually reinforcing precisely because they involve the same kind of reflective free play between imagination and understanding. The overlap and the epistemically relevant continuity between beauty and cognition resides in the fact that they both engage reflective judgment in an attempt to 'make sense', whether in the form of aesthetic or cognitive reflection.[24] Aesthetic contemplation is a reflection upon the various meanings of a representation, an attempt to make sense of it by trying out different interpretations, different ways of looking at it. This appears most clearly in what Guyer calls 'multicognitive' interpretations of the aesthetic free play as involving 'a multitude of concepts playfully applied to [a representation]', or 'occasioning [the understanding] to entertain fresh conceptual possibilities, while, conversely, the imagination, under the general direction of the understanding, strives to conceive new patterns of order'.[25] The imaginative exercise of the free play allows us to gain different insights into the object, and thereby beauty contributes directly to cognition.

While this interpretation is suggestive, it is based on the claim that cognitive and aesthetic reflective judgments involve the same free play between imagination and understanding – or at least that they are similar enough that they overlap in useful ways, so that one can contribute to the other. Yet, in my interpretation, they can't contribute to each other in the way that this interpretation suggests, since they consist in different mental processes. As I argued in Section 1, cognitive reflection is neither free nor harmonious, while aesthetic reflection is both free and harmonious.[26] Thus, if beauty is to contribute to cognition in any way, it cannot be on the basis of the epistemic contribution of aesthetic reflection.

Since the rational model of cognition entails that aesthetic pleasure cannot have any cognitive function, it seems that Kant's account simply rules out the idea of an epistemic contribution of beauty. However, the aim of the next section is to argue that it doesn't. As I will show, first, contrary to the rational model of cognition, feelings don't always obstruct or intrude upon cognition, they sometimes enhance it. Second, and more importantly for my purposes, feelings of beauty never obstruct or intrude upon cognition; rather, they always boost it. To support these claims, I turn to Kant's account of the effects of cognition on the faculty of feeling.

5 The aesthetic dimension of cognition

Our nature as knowers entails that our cognitive activity, and the cognitions that result from this activity, 'affects our feeling (by means of pleasure or displeasure)' (LL 34 [24:48]), and these effects are part of our cognitive life – they belong to what Kant calls the 'aesthetic perfection of cognition'. Unfortunately, he doesn't present a unified account of it, so some reconstructive work is called for.

Throughout the *Lectures on Logic*, Kant describes aesthetic cognition in a number of ways. An aesthetically successful cognition pleases the senses; it provides 'insight'; it agrees with 'our feeling and our taste'; it 'excites, delights and flatters our feeling'. It can be stimulating, attractive, exciting, lively. Aesthetic perfection is defined as 'new, easy, lively', resting on 'agreement with the subject' and 'the particular laws of human sensibility'. What is particularly relevant to the purpose of my argument is that aesthetic cognition encompasses a variety of feelings of pleasure, including aesthetic feelings: it involves 'real, independent beauty', 'the essentially beautiful' as well as merely agreeable feelings – what Kant calls the 'pleasant'.[27]

> Through this agreement with the universal laws of sensibility *the really, independently beautiful*, whose essence consists in *mere form*, is distinguished in kind from the *pleasant*, which pleases merely in sensation through stimulation or excitement, and which on this account can only be the ground of a merely private pleasure.
>
> (LL 547 [9:36-7])

To make sense of the contribution of feelings of beauty to cognition, it is essential to distinguish it from the contribution of pleasant feelings to cognition. Kant's lecture notes state that the pleasant belongs to the matter of sensibility. It pleases the senses in sensation, and can spoil logical perfection. In contrast, the essentially beautiful belongs to the form of sensibility. It consists in the agreement of a cognition with the laws of intuition, and combines best with logical perfection. The former is stimulating and exciting, while the latter is 'the object of a universal pleasure' (LL 547 [9:37]).[28] Let me examine them in turn in order to identify and spell out the distinctive features of the beautiful dimension of the aesthetic perfection of cognition.

Pleasant feelings may be conducive to the success of cognition in some cases. First, they can help detect salient features of the object and thereby facilitate the picking out of certain properties or patterns that we may not have detected at the logical or the conceptual level alone and that may be epistemically significant (e.g., parts of a scientific image that have a particularly nice colour may stand out). Second, they engage our capacity for attention by enlivening our mind and keeping the object in mind: 'Gentle excitement can give occasion for further reflection, to be sure' (LL 267-8 [24:808]). Yet even when pleasant feelings are helpful to cognition, their contribution to it is both extrinsic and contingent. First, they are merely instrumental to the pursuit of our cognitive ends. They are neither necessary nor sufficient for it, and they can even be counterproductive, as already noted in the case of obstruction. For instance, excitement and delight are unreliably helpful: 'stimulation and excitement, most of all, can spoil the logical perfection in our cognitions and judgments' (LL 547 [9:37]). Second, we have no epistemic justification for paying more attention to pleasant features than to indifferent or unhelpful ones. As merely subjective feelings, pleasant feelings are not sharable, and thus intrude upon cognition if used as if there were objective grounds for it.[29] Therefore, pleasant feelings are not reliably advantageous to cognition and in particular to its logical perfection. They always retain the potential to obstruct it and intrude on it.[30]

In contrast, Kant believes that 'aesthetic perfection in regard to the essentially beautiful can … be advantageous to logical perfection' (LL 547 [9:37]). To make sense of this claim, note that he repeatedly talks about beauty as 'the feeling … which animates (*Belebung*) the cognitive faculties' and 'indirectly … serve[s] cognition too' (CJ 194 [5:316]; translation modified).[31] Through the notions of animation, quickening and enlivening, he puts forward the idea that the stimulation of our cognitive faculties is conducive to their activity. By doing so, experiences of beauty in general enhance our cognitive activity and are thus advantageous to cognition. In other words, the more I experience beauty the better for my cognitive enhancement. Moreover, and more importantly for my purposes in my reading of Kant, experiencing a particular cognition as beautiful is good for my cognition of it. This is due to the fact that there is an intrinsic connection between the feeling of beauty and the efficiency of my cognitive activity. The experience of a cognition as beautiful stimulates the activity of imagination and understanding, and thus stimulates our cognitive activity as it occurs. For as argued in Section 1, cognitive and aesthetic activity can take place concomitantly as I experience a beautiful cognition. The feeling of beauty is thus a cognitive booster. By making us more efficient cognisers, it contributes to the logical perfection of a cognition.

Note, however, that the epistemic advantage afforded by the feeling of beauty only contributes to cognition indirectly, for it does so irrespective of whether the cognition is true or not. The intrinsic connection between beauty and cognition is not between beauty and epistemic credence or beauty and truth, but between beauty and the state of the cognitive faculties. The effect of the feeling of aesthetic pleasure is on the activity of cognising rather than the content of cognition. Beauty is a reliable sign of cognitive efficacy only insofar as it enhances cognitive activity as it occurs. What it tracks is the condition of our cognitive powers.

As a result, contrary to the rationalist model of cognition presented in Section 3, the feeling of beauty doesn't intrude upon cognition; it only impacts our cognitive powers. Nor does it obstruct cognition, since it only ever enlivens our cognitive faculties. However, in line with the rationalist model of cognition, the beauty of a cognition should not be taken to provide epistemic guidance. Nor should it be used as a means to choose between competing theories. For on my interpretation, beauty is not an indication of insight.[32] But while it cannot be used to ground epistemic choices, I have shown that it has a legitimate epistemic function, namely that of enhancing our cognitive powers.

On this basis, we can now make sense of why Kant repeatedly stresses that 'we must make it our task to provide aesthetic perfection for those cognitions that are in general capable of it, and to make a scholastically correct, logically perfect cognition popular through its aesthetic form' (LL 548 [37-8]). Human cognition is embodied, and through the demand for the aesthetic perfection of cognition, he not only acknowledges this but makes allowances for it: 'Aesthetic perfection consists in the agreement of cognition with the subject and is grounded on the particular sensibility of man' (LL 547 [9:36]). Of course, the aesthetic perfection of cognition is not necessary for cognition to be successfully pursued. But it certainly makes its pursuit both more enjoyable and more efficient. As imperfect, finite knowers with limited computational powers, we need all the cognitive help we can get, including from our aesthetic capacities. As Kant writes, 'the needs of human nature and the end of popularity in cognition demand, however, that we seek to unite the two perfections [logical and aesthetic] with one another' (LL 548 [9:37]). Although few cognitions may turn out to be beautiful, what I have shown is that they can, and in fact should, be beautiful.[33]

6 Conclusion

As mentioned at the beginning of this paper, a number of passages from Kant's *Lectures on Logic* as well as the third Critique suggest that for Kant, cognition cannot be beautiful.[34] What I set out to show is that his view of the relationship between cognition and beauty is not as straightforward as these passages seem to imply, and that we should think of beauty as connected to cognition in a number of important and meaningful ways.

Rather than summarise the various claims defended in this paper, I would like to conclude by taking stock of the notion of an aesthetic of cognition as I have defended it. I believe that it is an important addition to traditional ways of interpreting Kant. Kant's account of cognition is often characterised as 'impoverished', reduced to 'the acts of conceptual subsumption', as the mere conceptualisation of particulars, with a 'sharp divide between the aesthetic and the cognitive' (Pillow, 2006: 246, 248, 254). Longuenesse, for instance, talks of the divide between the cognitive work of determinative understanding and the merely reflective play of aesthetic experience (Longuenesse 1998: 164). In contrast with this view, the line I have defended here suggests a broader, and potentially richer, conception of cognition, a conception that brings together a wide array of cognitive processes that goes well beyond the mere determinative work of the understanding to include our aesthetic capacities. Thereby, I hope to have demonstrated that Kant's account of cognition is far from portraying human beings as disembodied, pure minds.[35]

Acknowledgements

I presented earlier drafts of this paper at the Visiting Speaker Seminar at the University of Sheffield, Kant's Scots in Edinburgh, the London Aesthetics Forum, and 'Shaping the Trading Zone: Bringing Aesthetics and Philosophy of Science Together' at the University of Leeds. I would like to thank all the participants for their feedback as well as Angela Breitenbach, Cain Todd and Christian Wenzel for their invaluable suggestions. Finally, special thanks go to Yoon Choi for her tireless encouragements and insightful comments.

Notes

1 As the following works by Kant are cited frequently, I have used the following abbreviations throughout the paper: LL (*Lectures on Logic*), CJ (*Critique of the Power of Judgment*), CPR (*Critique of Pure Reason*). The second reference is to the Akademie edition of Kant's works, using the translations from the Cambridge Edition of Kant's Works (Cambridge University Press).

2 See also 'There is neither a science of the beautiful, only a critique, nor beautiful science, only beautiful art' (CJ 184 5:304). 'No judgment at all can be made, however, concerning a beautiful cognition' (LL 37 24:51). 'No science can be beautiful' (LL 270 24:811). In this paper, I will not tackle the issue of the possibility of a science of the beautiful.

3 See for instance Rueger (1997), Koriako (1999), Wenzel (2001).

4 See for instance Shier (1998) and Wenzel (1999). In contrast with this view, I have argued elsewhere that Kant's account allows for both the aesthetically indifferent and the aesthetically ugly (Cohen (2013)).

5 For instance, Rueger has noted that 'the aesthetic pleasure that is characteristic of the free play of our faculties in reflective judgement is not to be found in the exercise of determinative judgement in science' (Rueger, 1997: 315).

6 See for instance Kant (1999): 155 A19/B33.

7 Of course, the expression 'at any given time' is infelicitous since our faculties are not meant to operate in time. But it is a figure of speech, as should be clear from the context of the discussion.

8 For instance, as Rueger notes, 'Science aims at the general under which the particular can be subsumed; only when this aim has been reached can we speak of science in the proper sense. For Kant, the title of a science could not be given to an activity – the aesthetic experience – that involves an unending vacillation between the particular and the general where the general is never found to adequately subsume the particular' (Rueger, (1997): 315).

9 See for instance CJ 175 5:295.

10 See also 'The attainment of every aim is combined with the feeling of pleasure' (CJ 73 5:187). This includes the aims of cognition. For a discussion of Kant's claim that this pleasure goes unnoticed, see Merritt (2014).

11 Allison puts forward a gradual model according to which there can be more or less harmony between the cognitive faculties rather than different kinds of harmony (Allison, (2001): 117). On Guyer's interpretation, aesthetic harmony is 'an *excess* of felt unity or harmony' (Guyer, (2005): 149). In contrast, on my reading, cognitive and aesthetic harmony differ in kind.

12 For an account of the relationship between harmony and pleasure, see Cohen (2017).

13 Ginsborg has the best formulation of the worry: 'if we regard the relation of the faculties in their free play as unique to aesthetic experience, it seems that Kant has no right to argue from the universal validity of empirical cognition to the universal validity of pleasure. For, … the free play of the faculties no longer appears to manifest a condition required for cognition' (Ginsborg, 2015: 54). I would like to thank Angela Breitenbach and Yoon Choi for pressing this challenge.

14 See CJ 124-5 5:240-1.

15 'Sofern sie unter einander, wie es zu einem Erkenntnisse überhaupt erforderlich ist, zusammen stimmen' (CJ 5:218). Pluhar's translation is even clearer on this point: 'this subjective universal communicability can be nothing but that of the mental state in which we are when imagination and understanding are in free play (insofar as they harmonize with each other as required for *cognition in general*)' (CJ 62 5:218). For contrasting interpretations of this passage, see for instance Guyer (1997): 85–6 and Ginsborg (2015): 92–3. However, my interpretation concurs with Ginsborg's conclusion that no free play is required for any act of cognition.

16 One may object that some cognitions, and scientific ones in particular, differ from roses insofar as the latter are sensible objects whereas the former are not. This would suggest that abstract scientific cognitions pose a particular problem for an account of the possibility of beautiful cognition. While it falls beyond the remit of this paper to discuss this issue, one could conjecture that my account could read as implying that not only can we find scientific objects beautiful while we cognise them, we can also find scientific cognitions beautiful. For a cognition to be judged beautiful, all that is necessary is that the judgment needs to be neither grounded on concepts nor aimed at concepts, but

rather grounded on the subject's feeling of disinterested pleasure. As long as judgment operates in relation to the faculty of pleasure alone, it is aesthetic, and remains so even if the representation that is apprehended is fully conceptualised and determined, as Kant hints at in CJ 90 5:204. Thus, irrespective of the type of representation we are faced with, as long as our judgment is based on disinterested pleasure, it is a judgment of taste. The only relevant factor is what is 'abstracted from the concept in his judgment' (CJ 115 5:231). Unfortunately, there is no space to defend this claim here. For enlightening discussions of the particular case of mathematical objects, see Breitenbach (2013) and Wenzel (2001). For an account that focuses on scientific representations in general, see Breitenbach (forthcoming). For a compelling attempt to show that Kant's philosophy has the resources to deal with cases of beautiful mathematics in spite of the fact that he himself may not have seen it, see Wenzel (2013).

17 For instance, 'In most discussions of the relations between emotion and cognition, the emphasis has been on the assumption that the former distorts the latter. For Kant, emotion was an illness of the mind' (Frijda, Mastead and Bem, 2000: 2).

18 See McAllister, 1996: 9ff.

19 'Representations can also be related to something other than cognition, namely, to the feeling of pleasure and displeasure (the way in which we are affected by things)' (LL 440 24:701). 'Feelings can never produce a cognition' (LL 466 24:730). 'Through sensation, good feeling, pain – one does not cognize an object' (LL 348 24:904).

20 'Through these we are transposed into a condition most unsuitable for judging' (LL 297 24:842).

21 We could draw an analogy between these dangerous consequences and the way in which children can interfere with adult conversation. They can make a lot of noise, which makes the conversation difficult to follow for the adults, putting some of them in a bad mood and thereby making the conversation less pleasant for them, or taking their attention away from the conversation (first type of damage: obstruction model). Or they can intervene in the conversation, expressing an opinion, raising concerns, etc. (second type of damage: intrusion model).

22 See also LL 157 24:198 and 125 24:158.

23 This is what Kant calls 'prejudice': 'The principal sources of prejudices are subjective causes, accordingly, which are falsely held to be objective grounds' (LL 315-6 24:864-5). A prejudice is a subjective ground that has been turned into an epistemic principle and thereby illegitimately plays the role of objective ground. This is why, for Kant, the cause of error cannot be found in the understanding itself: 'the understanding taken alone cannot possibly err' (LL 64 24:84).

24 See in particular Wenzel (2013), esp. 63, 67, Breitenbach (2013), esp. 90–3 and (2015), esp. 11–5, and for a more contemporary formulation of this point, Elgin (2002). On Breitenbach's account, the beauty of mathematics resides in the spontaneous activity of our conceptual and imaginative capacities. While her account is focussed on the particular case of mathematical knowledge, I believe that it can be extended to cognition in general.

25 Respectively in Guyer (2006): 166, Seel (1988): 344–9 and Allison (2001): 171. On Guyer's account for instance, 'a beautiful object suggests an indeterminate or open-ended manifold of concepts for the manifold of intuition, allowing the mind to flip back and forth playfully and enjoyably among different ways of conceiving the same object without allowing or requiring it to settle down on one determinate way of conceiving the object. … the free play is precisely among a multiplicity of possible concepts and hence cognitions suggested by the beautiful object' (Guyer, 2006: 166).

26 Keren Gorodeisky has explored another way of distinguishing between beauty and cognition. As she has noted, 'the patterns of order apprehended and exhibited by the imagination in judgments of taste differ in kind from those apprehended and exhibited in cognitive judgments' (Gorodeisky, 2011: 420). I believe that our accounts are compatible but focus on different aspects of Kant's account – I focus on its affective dimension while she focuses on its imaginative dimension.

27 Respectively in LL 24 9:37, 186 24:234, 39 9:54, 44 9:60, 266 24:806, 547 9:36–7.

28 See also LL 266–71 806–12, 443–6 705–9, 546–8 36–8.

29 'Such a pleasure would be none other than mere agreeableness in sensation, and hence by its very nature could have only private validity, since it would immediately depend on the representation through which the object *is given*' (CJ 102 5:217). In contrast, objective grounds 'are independent of the nature and interest of the subject' (LL 574 [9:70]). They can be adopted by all, at least in principle: they are 'valid for the reason of every human being to take it to be true; ... regardless of the difference among the subjects' (CPR 685 [A820–1/B848–9]).

30 One may worry that on my account, cognitive pleasures, which I defined in Section 1 as the pleasures we feel when we cognise, turn out to be a liability for cognition, just as pleasant feelings discussed above do. However, they don't. For in contrast with pleasant feelings, cognitive pleasures are the effects on feeling of a particular type of interplay between imagination and understanding, namely their unfree harmony. Thus, just as with beauty, it is insofar as it is harmonious that cognition is pleasurable. However, in contrast with aesthetic pleasure, which enlivens the activity of the cognitive faculties in a noticeable and ongoing fashion, cognitive pleasures are discreet and barely noticeable since they are so ubiquitous (see e.g. CJ 74 5:187). Insofar as they are triggered by the attainment of our cognitive aims, their noticeable effects are essentially motivational (i.e., they engage our cognitive interest and thus our drive for further cognitions). I would like to thank Yoon Choi for pressing me to address this worry.

31 'When the aim is aesthetic, then the imagination is free, so that, over and above that harmony with the concept, it may supply, in an unstudied way, a wealth of undeveloped material for the understanding which the latter disregarded in its concept. But the understanding employs this material not so much objectively, for cognition, as subjectively, namely to *quicken (Belebung) the cognitive powers*, though indirectly *this does serve cognition too*' (CJ 185 5:316-7; my emphasis). This is Pluhar's translation. See also CJ 107 5:222 where Kant talks about beauty's 'internal causality (which is purposive) with regard to cognition in general'. This is something Zuckert briefly hints at although she doesn't account for the way in which the liveliness of our faculties helps us as cognisers (Zuckert, 2007: 453). For other occurrences of the notion of *Belebung*, see CJ 104 5:219, 107 5:222, 122 5:238, 167 5:287, 192 5:313, 193 5:315, 206 5:329, 207 5:331.

32 Contrast with Breitenbach's claims that 'beauty can provide a heuristic means for choosing between theories, even though there is no intrinsic connection between beauty and truth' and that 'Aesthetic considerations may therefore provide an initial, even if not determining, indicative guide in our search for understanding of the phenomena' (Breitenbach, 2013: 94, 96). I would like to thank her for helping me pinpoint where our disagreement lays.

33 As Kant notes, 'History, geography, reading the ancients, which unite both perfections, anthropology [too], must be our instructors and must make the spirit more alert' (LL 270 24:811). 'There always remains a kind of conflict between the aesthetic and the logical perfection of our cognition, which cannot be fully removed' (LL 547–8 9:37).

34 See footnote 2.

35 Contrast with 'in the veins of the knowing subject, such as . . . Kant [has] construed him, flows not real blood but rather the thinned fluid of reason as pure thought activity' (Dilthey, 1922: viii). For an account of other features that are due to human being's embodied cognition, see Cohen (2014).

References

Allison, H. (2001). *Kant's Theory of Taste: A Reading of the Critique of Aesthetic Judgment*. Cambridge: Cambridge University Press.

Breitenbach, A. (2013). 'Aesthetics in Science: A Kantian Proposal'. *Proceedings of the Aristotelian Society*, 113: 83–100.

Breitenbach, A. (2015). 'Beauty in Proofs: Kant on Aesthetics in Mathematics'. *European Journal of Philosophy*, 23: 955–77.

Breitenbach, A. (forthcoming). 'The Beauty of Science without the Science of Beauty'. *Journal of the History of Philosophy*.

Cohen, A. (2013). 'Kant on the Possibility of Ugliness'. *British Journal of Aesthetics*, 53 (2): 199–209.

Cohen, A. (2014). 'The Anthropology of Cognition and its Pragmatic Implications', in A. Cohen (ed.) *Kant's Lectures on Anthropology: A Critical Guide*. Cambridge: Cambridge University Press: 76–93.

Cohen, A. (2017). 'Kant on Emotions, Feelings and Affectivity', in M. Altman (ed.) *The Palgrave Kant Handbook*. London: Palgrave Macmillan, forthcoming.

Dilthey, W. (1922). *Einleitung in Die Geisteswissenschaften Versuch Einer Grundlegung Für Das Stidium der Gesellschaft Und der Geschichte*. Leipzig: Teubner.

Elgin, C. (2002). 'Art in the Advancement of Understanding'. *American Philosophical Quarterly*, 39 (1): 1–12.

Frijda, N. H., Mastead, A. S. R. and Bem, S. (2000). *Emotions and Beliefs: How Feelings Influence Thoughts*. Cambridge, UK: Cambridge University Press.

Ginsborg, H. (2015). *The Normativity of Nature: Essays on Kant's Critique of Judgment*. Oxford: Oxford University Press.

Gorodeisky, K. (2011). 'A Tale of Two Faculties'. *British Journal of Aesthetics*, 51 (4): 415–36.

Guyer, P. (1997). *Kant and the Claims of Taste*. Cambridge: Cambridge University Press.

Guyer, P. (2005). *Values of Beauty: Historical Essays in Aesthetics*. Cambridge: Cambridge University Press.

Guyer, P. (2006). 'The Harmony of the Faculties Revisited', in R. Kukla (ed.), *Aesthetics and Cognition in Kant's Critical Philosophy*. Cambridge: Cambridge University Press.

Kant, I. (1999). *Critique of Pure Reason*. Cambridge: Cambridge University Press.

Koriako, D. (1999). *Kants Philosophie der Mathematik: Grundlagen, Voraussetzungen, Probleme*. Hamburg: Meiner.

Longuenesse, B. (1998). *Kant and the Capacity to Judge: Sensibility and Discursivity in the Transcendental Analytic of the Critique of Pure Reason*. Princeton, MA: Princeton University Press.

McAllister, J. W. (1996). *Beauty and Revolution in Science*. Ithaca/ London: Cornell University Press.

Merritt, M. (2014). 'Kant on the Pleasures of Understanding', in A. Cohen (ed.), *Kant on Emotion and Value*. Basingstoke: Palgrave Macmillan: 126–145.

Pillow, K. (2006). 'Understanding Aestheticized', in R. Kukla (ed.), *Aesthetics and Cognition in Kant's Critical Philosophy*. Cambridge: Cambridge University Press.

Rueger, A. (1997). 'Experiments, Nature and Aesthetic Experience in the Eighteenth Century'. *British Journal of Aesthetics*, 37: 305–22.

Seel, G. (1988). 'Uber den Grund der Lust an schönen Gegenständen: Kritische Fragen an die Asthetik Kants', in Hariolf Oberer and Gerhard Seel (eds.), *Kant: Analysen-Probleme-Kritik*. Wiirzburg: Konigshausen and Neumann, 317–56.

Shier, D. (1998). 'Why Kant Finds Nothing Ugly'. *British Journal of Aesthetics*, 38 (4): 412-18.

Wenzel, H. C. (1999) 'Kant Finds Nothing Ugly?' *British Journal of Aesthetics*, 39 (4): 416–22.

Wenzel, H. C. (2001). 'Beauty, Genius, and Mathematics: Why Did Kant Change His Mind?' *History of Philosophy Quarterly*, 18: 415–32.

Wenzel, H. C. (2013). 'Art and Imagination in Mathematics', in Michael L. Thompson (ed.) *Imagination in Kant's Critical Philosophy*. Berlin: Walter de Gruyter, 49–68.

Zuckert, R. (2007). 'Kant's Rationalist Aesthetics'. *Kant-Studien*, 98 (4): 443–63.

9 Epistemology as fiction

Adam Toon

1 Introduction

In *A Neurocomputational Perspective*, Paul Churchland poses a fundamental and far-reaching challenge to epistemology and philosophy of science. In their descriptions of our epistemic practices, both epistemologists and philosophers of science typically rely upon our ordinary categories for describing the operations of the mind – in particular, the notion of *belief* – commonly described as *folk psychology*. In contrast, Churchland's *eliminative materialism* holds that folk psychology is a bad theory that will be replaced by the latest theories in cognitive science (Churchland, 1981). As well as requiring a radical shift in our conception of the nature of the mind, Churchland argues that eliminative materialism also undermines the theories of traditional epistemology and philosophy of science. Philosophers of science in particular must abandon their existing 'sentential' framework for discussing scientists' reasoning, knowledge and understanding. In its place, they must learn to adopt a radically new framework, in which the central notions are drawn from the technical vocabulary of neural network modelling (Churchland, 1989).

Churchland's analysis focuses on the brain. By contrast, much recent work in cognitive science stresses the importance of interaction of brain, body and world in carrying out cognitive tasks (e.g. Robbins and Aydede, 2008). Familiar examples include the way that we gesture when reasoning or reach for pen and paper when trying to solve a crossword puzzle. In light of this work, some authors have argued that cognition, and even mental states, sometimes *extend* beyond the brain and body into the world (Clark and Chalmers, 1998; Clark, 2008). In previous work, I have argued that the extended mind thesis has important implications for the way that we understand scientists' cognitive activity (Toon, 2014, 2015; for related approaches, see Bechtel, 1996; Nersessian, 2005; Giere, 2006). From this perspective, Churchland's favoured successor to folk psychology looks too narrow in its focus on the brain alone. And yet, his eliminativist challenge to philosophy of science must still be faced. Even if our best theory of cognition encompasses not only the brain, but also interaction between brain, body and environment, it might still turn out to stand at odds with the traditional, sentential framework of folk psychology.

How should we respond to this challenge? Eliminativism is a hard road to follow. Folk psychology is intricately woven into the fabric of our language, and into the way that we talk about reasoning, knowledge and understanding in particular. Embracing eliminativism would therefore require a radical transformation in the way that we try to make sense of people and their epistemic activities. Moreover, given the difficulties faced by Churchland's own approach, it remains far from clear what framework we ought to adopt in its place. This paper explores an alternative response to the threat of eliminativism. *Mental fictionalism* is the view that even though folk psychological states might not exist it is useful to talk as if they do. According to the fictionalist, mental states, like beliefs and desires, are useful fictions (for discussion, see Wallace, 2007; Demeter, 2013a). In previous work, I have argued that a fictionalist approach captures the spirit of much of our ordinary talk about the mind (Toon, 2016). In this paper, I will argue that fictionalism also provides philosophers of science with a promising way to respond to Churchland.

The discussion will proceed as follows. In Section 2, I will introduce eliminativism, focusing on Churchland's account of the nature of understanding. In Section 3, I will argue that Churchland's account is too restrictive and that understanding is often an extended mental state. In Section 4, I will introduce mental fictionalism. Finally, in Section 5, I will show how fictionalism allows philosophers of science to respond to the threat of eliminativism.

2 Eliminativism and epistemology

A key feature of folk psychology is the attribution of propositional attitudes. We say that Barbara believes that planes can fly, that George wants to go to the cinema, that Adam hopes that Derby County will win promotion this year, and so on. Such talk is also central to epistemology. To know that *p* is to have the belief that *p*, where that belief is true and suitably justified (or obtained in the right sort of way, etc.). Reasoning is taken to involve moving from one set of claims (e.g. that the body was found in the library, that the only person with a key was the butler) to another (e.g. that the butler did it). Classical approaches in cognitive science hope to find beliefs, desires and other propositional attitudes as language-like structures inside the head (e.g. Fodor, 1975). By contrast, Churchland argues that 'the sentential kinematics of folk psychology is but a commonsense *theory*, and almost certainly a *false* theory' (1989, p. xvi). In fact, our best theories in cognitive science reveal that 'the basic kinematics of cognitive creatures is a kinematic not of sentences but of high-dimensional activation vectors being transformed into other such vectors by passing through large arrays of synaptic connections' (1989, p. xvi). As a result, we ought to stop talking about beliefs, desires and other propositional attitudes and instead adopt the language of our best theories of mind and cognition.

Like epistemologists in general, philosophers of science also tend to use the 'sentential' framework of folk psychology when describing scientists' reasoning, knowledge and understanding.[1] As a result, they too face the challenge posed by eliminativism. Let us focus on understanding in particular, since it is here that Churchland develops his approach in most detail. Churchland motives his account by pointing to the shortcomings of the deductive-nomological model of explanation (DN model) (Hempel, 1965). According to the DN model, we explain a phenomenon by showing how it may be deduced from some general law, together with relevant initial conditions. While the conceptual difficulties of the DN model are well-known, Churchland argues that it is also psychologically unrealistic. People often possess explanatory understanding without being able to articulate the required laws or initial conditions. Moreover, they are often unable to perform the necessary deductions, at least not with the swiftness with which understanding often dawns. According to Churchland, such difficulties are not specific to the DN model. Instead, they stem from 'the fundamental assumption that language-like structures of some kind constitute the basic or most important form of representation in cognitive creatures, and the correlative assumption that cognition consists in the manipulation of those representations by means of structure-sensitive rules' (1989, p. 154).

As William Lycan (1996) has pointed out, Churchland's criticism of the DN model is somewhat unfair, since most of its proponents did not aim to describe scientists' cognitive processes (see also Bechtel, 1996). Such authors tended to focus on *explanation*, rather than *understanding*: while accounts like the DN model sought to spell out what makes a good scientific explanation, understanding was taken to be a subjective, psychological phenomenon of little interest to philosophers of science. Times have changed, however. Among both epistemologists and philosophers of science, there is now widespread agreement that understanding is an important cognitive state that we ought to try to analyse (Kvanvig, 2003; de Regt, *et al.*, 2009). It is also commonly agreed that understanding goes beyond merely knowing the various facts and theoretical principles needed to explain some phenomenon. To possess understanding, it is said, someone must also 'see' or 'grasp' how the various principles and facts fit together. In this vein, Wayne Riggs (2003, p. 218) writes that:

> [a]n important difference between merely believing a bunch of true statements within subject matter M, and having understanding of M, is that one somehow sees the way things fit together. There is a pattern discerned within all the individual bits of information or knowledge.

Consider the question 'why do planes fly?' On this view of understanding, someone who understands why planes fly can do more than simply recall Bernoulli's principle, recite facts about air pressure and so on; she also 'grasps' or 'sees' the connections between these things. For example, she must 'grasp' how Bernoulli's principle applies to the air flow around the wing, 'see' how the difference in air speed will result in a difference in pressure and so on. This conception of understanding differs from the DN model in a number of respects. In particular, it does not require that the relationship between the different propositions involved be deductive. And yet it remains resolutely 'sentential' in its approach and so, if Churchland is correct, must ultimately be discarded.

In place of the DN model and other sentential accounts of understanding, Churchland proposes his own *prototype-activation model* (PA model). The key theoretical notions underpinning the PA model are drawn from *artificial neural networks* or *connectionist* models of the brain. Put simply, connectionist networks are collections of input-output devices called *units*. Each unit can be activated to various degrees and is connected to other units by 'synaptic' connections with different *weights*. When the input units are activated, a *pattern of activation* spreads throughout the network, resulting in a particular output. The way that the network responds to a particular input is determined by the weights of the connections between its units. Once the network's connection weights are adjusted correctly, it will learn to categorise input patterns into different groups. One of Churchland's main examples is a network that learnt to categorise sonar echoes into two groups: those resulting from underwater mines and those resulting from rocks on the seabed. For each category that the network can distinguish, there will be a *prototypical response*. In the rock-mine network, there is the pattern of response corresponding to a prototypical mine or prototypical rock. Responses to actual mines or rocks will tend to cluster around these prototypes (Churchland, 1989, pp. 200–206).

According to Churchland, connectionism provides a 'possible conception of knowledge or understanding that owes nothing to the sentential categories of current common sense' (1989, p. 177). For example, an individual's *theory* of the world, Churchland argues,

> is not a large collection or a long list of stored symbolic items. Rather, it is a specific point in that individual's synaptic weight space. It is a configuration of connection weights, a configuration that partitions the system's activation-vector space(s) into useful divisions and subdivisions relative to the inputs typically fed the system.
>
> (1989, p. 177)

The connectionist framework is also central to Churchland's PA model of understanding. According to the PA model,

> understanding consists in the activation of a specific prototype vector in a well-trained network. It consists in the apprehension of the problematic case as an instance of a general type, *a type for which the creature has a detailed and well-informed representation*.
>
> (1989, p. 210)

For example, in the case of the rock-mine network, understanding might consist in the fact that, when presented with a sonar echo from a mine that it has not yet encountered, the network follows a pattern of activation close to that of the prototypical mine.

Churchland argues that abandoning the sentential framework will transform our approach to a whole raft of issues in philosophy of science, including 'the nature of theories, the theory-ladeness of perception, the nature of conceptual unification, the virtues of theoretical simplicity, the nature of paradigms, the kinematics of conceptual change, the character of abductive inference, and the nature of explanatory understanding' (1989, p. xv). And yet, as Churchland himself acknowledges, pursuing this new approach is far from straightforward. As we noted earlier, abandoning the categories of folk psychology would require a fundamental shift in the way that we talk about people and their behaviour – and their reasoning, knowledge and understanding in particular. For example, Churchland notes that when it comes to saying what it is that makes one explanation better than another, 'we must answer carefully, since we are denied the usual semantic vocabulary of reference, truth, consistency, entailment, and so forth' (1989, p. 220). Instead, we must learn to answer such questions within a radically different framework of networks, connection weights and activation vectors.

3 Extended cognition

Churchland's account focuses on the brain. In response, Bill Bechtel (1996) has argued that this approach is too narrow. In fact,

> Churchland is mistaken in localizing the focus of philosophy of science exclusively in activities occurring in the heads of scientists. While representations are central to scientific activity, the representations that matter are not exclusively mental representations. They are also external representations such as are found in sentences of natural language as well as in tables, figures, and diagrams.
>
> (1996, p. 122; for similar criticism of Churchland, see Giere, 2002)

External representations are not merely vehicles for communication, Bechtel argues. Instead, they play a key role in scientists' reasoning processes. For example, 'constructing an explanation is an interactive activity involving both the cognitive agent and various external representational systems' (Bechtel 1996, p. 126). To illustrate this idea, Bechtel cites an influential discussion of multiplication by the San Diego connectionist group (Rumelhart, *et al.*, 1986). Most of us aren't able to multiply three-digit numbers in our heads. If we're given pen and paper, however, the task becomes much easier: we can write the numbers down one underneath the other and work step-by-step through the procedure for long multiplication. In this way, we transform a complex task (e.g. multiplying 546 by 837) into a series of much simpler tasks (e.g. multiplying 3 by 4, remembering to carry the 1 and so on). While connectionist models might explain what is happening in our brain when we do this, the overall task is accomplished by a larger system that includes external, material representations.

In recent years, a growing body of work in cognitive science has revealed that much of our cognitive activity has a similar character, involving productive cooperation between internal and external resources (e.g. Robbins and Aydede, 2008). In light of this work, some philosophers have endorsed the *hypothesis of extended cognition* (HEC) (e.g. Clark and Chalmers, 1998; Clark, 2008; for related ideas, see Menary, 2007; Rowlands, 1999; Wheeler, 2005; and Wilson, 2004). According to HEC, external devices can sometimes become part of our cognitive processes. On this view, the pen and paper we use in long multiplication is not simply a useful tool; it is part of the mechanism that realises our cognitive processes, just like the neurons in our brain. Proponents of the *extended mind thesis* go further and argue that it is not only cognitive processes that can be realised by external devices; even mental states, such as beliefs, can extend outside the head. Thus, Clark and Chalmers (1998) offer the well-known example of Otto, an Alzheimer's patient who carries a notebook wherever he goes to record useful information. According to Clark and Chalmers, Otto's notebook plays a similar role to normal biological memory. As a result, they argue, we ought to count the entries in the notebook as Otto's (dispositional) beliefs. Only an unmotivated bias for skin and skull would lead us to deny this (for debate over the extended mind thesis, see Menary, 2010).

Bechtel's criticism focuses mainly on what we might call *explanatory inquiry* – that is, the reasoning process that scientists follow in order to construct a new explanation for some phenomenon (1996, pp. 131–5). At first glance, we might think that, even if explanatory inquiry involves external representations, still the *outcome* of that process – that is, the state of understanding itself – remains inside the head. Elsewhere, I have argued that this would be a mistake. In fact, understanding is often an extended mental state (Toon, 2015). To see this, suppose that, rather than 'why do planes fly?', we consider a more difficult question, like 'why do planes experience Dutch roll?' Dutch roll is an oscillatory motion that can affect planes when they fly through turbulence. It is a fairly complicated phenomenon: merely writing down the equations of motion required to explain it requires a page or two, and textbooks normally include a series of diagrams and graphs to show the sequence of steps in a typical Dutch roll cycle. Now suppose that Barbara is an aeronautical engineer. When she is asked why planes experience Dutch Roll, Barbara is able to write down the relevant theoretical principles and facts about air pressure, plane's wings and so on. She can also show how these principles and facts combine to lead to Dutch roll. Without pen and paper, however, Barbara is unable to work through these steps. I suggest that this is a case of extended understanding. Barbara understands Dutch roll: she not only knows the relevant facts and theoretical principles; she also 'sees' or 'grasps' the connections between them. It is simply that these acts of 'seeing' or 'grasping' don't happen entirely inside her head (for further discussion, see Toon, 2015).

Of course, Churchland is well aware that scientists make use of pen and paper. And yet, he insists, the use of external representational devices should be kept distinct from understanding proper:

> The prototype activation model is focused first and foremost on what it is to have explanatory understanding of a problematic thing, event, or state of affairs. The linguistic expression, exchange, or production of understanding, should there be any, is an entirely secondary matter.
>
> (1989, p. 198)

According to Churchland, an explanation – in the sense of a set of sentences written down on paper – may *represent* a scientist's understanding of some phenomenon, and it may be enough to *cause* that understanding in others (1996a, p. 258). The understanding itself, however, remains entirely internal. The notion of extended understanding suggests that this view of the role of external representations is mistaken. In Barbara's case, external representations do not serve merely to represent her understanding or pass that understanding on to other people. Instead, the pen and paper is itself part of the material basis that realises Barbara's understanding of the phenomenon. In criticising the DN model, Churchland objects that working through a DN explanation takes some time, whereas understanding often dawns upon us almost immediately. We see that the kitchen is filled with smoke and realise straightaway that the toast is burning (1989, p. 199). Barbara's case serves to remind us that not all understanding is like this, however. In the case of more complex phenomena, such as that of Dutch roll, exercising our grasp of the relevant facts may take more time and involve prolonged and highly skilled interaction with external, material devices.

If scientists' cognitive processes and mental states can extend into the environment, then Churchland's own approach looks too narrow. What about traditional, sentential epistemology? In some respects, it might now seem to be on safer ground. For even if Churchland is right to say that the brain does not operate on language-like entities, such entities *will* often be found in scientists' extended cognitive systems. After all, Barbara *writes down* the various facts and theoretical principles required to explain Dutch roll. In this sense, it might seem that the sentential story is vindicated (Bechtel, 1996; see also Clark, 1989).

And yet not all cases of understanding will involve external, linguistic representations. In this vein, Churchland (1996b) argues that non-human animals and pre-linguistic children can possess understanding. For Churchland, this supports his non-linguistic, PA model. Rather than insisting upon a single, universal account of the nature of understanding, however, we might instead allow for a range of different cases. In some instances, understanding might depend upon external, linguistic representations, while in others it might not. And, of course, even where external representations are involved, they might not be linguistic (Bechtel, 1996). Rather than writing down various facts and principles, for example, Barbara might instead resort to working with graphs and diagrams, sketching the plane from various angles and positions. Alternatively, she might need to pick up a plastic model of a plane, imagining it being buffeted by a crosswind and tilting its wings from side to side. Or she might gesture with her hands instead. Each of these cases might require a different account of the underlying dynamics of cognition and each might find itself at odds with a traditional, sentential story.

So it appears that, although the notion of extended cognition might point to shortcomings in Churchland's own approach, his eliminativist challenge must still be met. Can we continue to use the sentential framework of folk psychology and talk about scientists believing particular propositions, inferring from one claim to another and 'seeing' connections between them? Or must we search for a radical new framework – or perhaps frameworks – which can capture the complex interplay that takes place between scientists' internal, pattern-matching abilities, bodily skills and external, material devices? In response to this challenge, I suggest that we turn to fictionalism.

4 Mental fictionalism

According to the fictionalist, folk psychological states like beliefs and desires are useful fictions (Wallace, 2007; Demeter, 2013a). In earlier work, I have developed a version of this approach by drawing on Kendall Walton's theory of fiction (Toon, 2016). The guiding idea behind this approach is that our ordinary talk about mental states can be understood along the same lines as acts of pretence within a game of make-believe.

Suppose that some children are playing with a doll. In Walton's terminology, the doll is a *prop* and the rules that govern the children's game are called *principles of generation* (Walton, 1990). The props in a game, together with its principles of generation, require the children to imagine certain things. For example, if the doll is in its pushchair, then the children are to imagine that the baby is in its pushchair. In Walton's terminology, this is *fictional* in the game. Children also *participate* in games in various ways. For example, they might push the doll along in the pushchair, thereby making it fictional that they are taking the baby for a walk. They also participate verbally. Significantly, for our purposes, these acts of verbal participation can be used to make genuine assertions. Suppose George says, 'The baby's in her pushchair'. When he says this, George isn't *really* claiming that there is a baby in the pushchair; he is only pretending. And yet, by doing so, George indicates that pretending in this way is appropriate. As a result, George *does* make a genuine assertion: he claims that the state of the props is such that to pretend in the way that he does is, fictionally, to speak the truth. In other words, he claims that the doll is in the pushchair.

Games with dolls are what Walton (1993) calls *content oriented*: the children's interest is not with the doll *per se*, but with the content of the make-believe world it helps to create. In some cases, however, our interest lies in the props themselves; the role of make-believe is to help us to understand the props. Walton (1993) argues that this *prop-oriented* make-believe may be found in cases of metaphor:

> Where in Italy is the town of Crotone?, I ask. You explain that it is on the arch of the Italian boot. 'See the thundercloud over there – the big angry face near the horizon,' you say; 'it is headed this way'. Plumbers and electricians distinguish between 'male' and 'female' plumbing and electrical connections. [...]
>
> All of these cases are linked to make-believe. We think of Italy and the thundercloud as something like pictures. Italy (or a map of Italy) depicts a boot. The cloud is a prop which makes it fictional that there is an angry face. Male and female plumbing or electrical connections are understood

> to be, fictionally, male and female sexual organs. [...] But our interest, in these instances, is not in the make-believe itself, and it is not for the sake of games of make-believe that we regard these things as props. [...]
>
> Make-believe [...] is useful in these cases [...] for articulating, remembering, and communicating facts about the props – about the geography of Italy, or the identity of the storm cloud, or functional properties of plumbing or electrical fixtures [...]. It is by thinking of Italy or the thundercloud or plumbing connections as potential if not actual props that I understand where Crotone is, which cloud is the one being talked about, or whether one pipe can be connected to another.
>
> (Walton, 1993, pp. 40–41)

Suppose that Elaine says 'Crotone is on the arch of the Italian boot'. When she says this, Elaine is not claiming there really *is* a giant boot in the Mediterranean; she is involved in pretence. And yet, like George when he talks about the baby, Elaine also makes a genuine assertion: she claims that the state of the props is such that to pretend in the way that she does is, fictionally, to speak the truth. In other words, Elaine asserts that Crotone is in a particular spot on the southern coast of Italy. In this case, make-believe is not interesting for its own sake; instead, it provides a vivid and memorable way of communicating a fact about the props in the game, that is, about the geography of Italy.

I suggest that we understand ordinary talk about mental states along similar lines. The Italian boot provides a useful game for understanding the geography of Italy. In a similar manner, folk psychology provides a useful game for understanding people and their behaviour. In this game, we imagine that people have certain states inside their heads, such as beliefs and desires. We also imagine that these states are caused by certain experiences, interact in certain ways and cause certain sorts of behaviour. We are no more committed to the existence of this inner machinery than we are to the existence of the Italian boot. And yet, pretending that this machinery exists serves an important purpose, providing us with an enormously valuable means of explaining and predicting people's behaviour.

Although folk psychological talk involves pretence, it also allows us to make genuine assertions. Suppose that Elaine says, 'George wants to go to the cinema tonight'. According to the fictionalist, when she says this, Elaine is not claiming that there is a particular sort of causal state inside George's head. Instead, she is invoking a familiar pretence within the game of folk psychology. By doing so, however, Elaine does make a genuine assertion about George: she claims that he is in some state *S* such that, fictionally, she speaks the truth. There are many different situations that would make Elaine's pretence appropriate: George might be standing patiently in a queue outside the box office, eagerly buying his tickets online, loudly complaining that he has to work tonight instead and so on. Invoking the game of folk psychology provides Elaine with a concise and extremely valuable – in fact, indispensable – means for picking out this disparate set of scenarios, and thereby describing George and his behaviour.

In earlier work, I have tried to demonstrate the attractions of this way of understanding folk psychology and its advantages over related approaches, such as behaviourism and instrumentalism (Toon, 2016). In what follows, I want to focus on how fictionalism can help us to respond to the eliminativist's challenge to epistemology and philosophy of science.

5 Epistemology as fiction

The eliminativist claims that folk psychology is a bad theory of mind and cognition. As a result, talk of beliefs, desires and other mental states should go the same way as talk about witches and phlogiston. While it might be difficult to accept, says the eliminativist, this harsh lesson applies to philosophers of science no less than it does to the folk. They too must stop talking about scientists believing certain claims, inferring from one proposition to the next and so on. This 'sentential' view of cognition must be jettisoned as a faulty picture of the reality of scientists' cognitive lives. In its place, we must learn to adopt the radical new theories formulated by our best cognitive science. For Churchland, this means embracing the technical vocabulary of connectionism and learning to describe and assess scientists' reasoning, knowledge and understanding in terms of networks, activation vectors and prototypes. As we have seen, more recent movements in cognitive science point in a different direction, seeking to capture complex interactions between our brains, bodies and external devices – some linguistic, others not. Taking this route, the final framework(s) that might replace folk psychology remain as yet unclear.

Fictionalism suggests an alternative path. While the eliminativist claims that folk psychology is a bad theory, the fictionalist denies that the folk were trying to give a theory of mind and cognition in the first place. According to fictionalism, when the folk say that someone has a particular belief, they are not making a claim about their inner machinery. Instead, talk about mental states is a useful fiction. As a result, the legitimacy of folk psychology does not depend upon cognitive science discovering beliefs and desires inside the head. Even if such states do not exist, says the fictionalist, it is useful to talk as if they do. This suggests a way for philosophers of science to respond to Churchland's challenge. Rather than treating the traditional, sentential framework as a theory of the nature of cognition, we ought instead to regard it as a useful metaphor for describing scientists' reasoning, knowledge and understanding. Crucially, I suggest, this metaphor serves to encompass not only what happens inside the scientists' head, but also their interactions with external, material devices (for a similar position, see Dennett, 1996).

To see how this might work, let us return to the case of understanding. As we have seen, both epistemologists and philosophers of science typically characterise understanding in sentential terms. To understand a phenomenon, it is said, someone must believe a set of propositions about it and 'see' or 'grasp' the connections between them. Churchland urges us to reject this view as a mistaken attempt to capture the underlying dynamics of cognition. The fictionalist suggests that we resist this interpretation of folk discourse. When we attribute understanding to someone, we are not claiming that they have a set of language-like structures inside their head. Instead, we are invoking the game of folk psychology; we are availing ourselves of a familiar and invaluable form of pretence. As a result, the epistemologists' characterisation of understanding in sentential terms need not be given up if it turns out that the brain does not operate on language-like structures.

When we invoke our pretence, we do make a genuine assertion: we claim that the person to whom we are attributing understanding is in some state *S* such that, fictionally, we speak the truth. Importantly, this state can encompass factors that extend beyond the person's brain and body. As we saw in Section 3, understanding can take any number of different forms and will often involve interaction with external devices. Barbara's understanding of Dutch roll might depend upon her writing down equations with pen and paper, drawing diagrams, working with models or even gesturing. When we say that Barbara understands Dutch roll, we are saying that she is in any one of *these* states – states such that it is appropriate to pretend as we do. In some of these cases, there will be language-like structures involved, so that our sentential characterisation of understanding might come close to the truth. In other cases, the underlying dynamics of cognition might stand at odds with the sentential framework. Despite this, each of these different scenarios counts as a case in which our pretence is appropriate.

Why the detour via make-believe? Why talk about things that don't exist? One reason is that metaphor allows us to express things that we cannot express in a straightforward, literal description. Recall the Italian boot. In this case, we can offer a literal paraphrase for our assertion about Crotone: we are claiming that it lies on the southern coast of Italy between Capo Colonna and Taranto. In other cases, however, a literal paraphrase might not be available. Thus, Stephen Yablo argues that some metaphors are *representationally essential* (1998, p. 250). In such cases,

> the language might have no more to offer in the way of a unifying principle for the worlds in a given content than that they are the ones making the relevant sentence fictional. It seems at least an open question, for example, whether clouds we call *angry* are the ones that are literally F, for any F other than 'such that it would be natural and proper to regard them as angry if one were going to attribute emotions to clouds'.
>
> (Yablo, 1998, p. 250)

Like the metaphor of angry clouds, our folk psychological metaphors might also be representationally essential (Toon, 2016). In some cases, understanding might involve manipulating external, linguistic structures. In others, it might involve diagrams, graphs or models. In still other cases, it might be entirely internal. Despite the hopes of Churchland and others to offer a single, overarching theory of understanding, it might turn out that there is little in common between these various different cases, apart from the fact that each counts as a case in which it is appropriate to attribute understanding within the game of folk psychology.

Metaphors also bring further benefits. They introduce 'framing effects' in which we are asked to 'see' our primary subject (e.g. Italy, clouds) in terms of another, secondary subject (e.g. a boot, emotions) (Moran, 1989; Beardsley, 1962). This can be extremely fruitful, leading us to a host of further insights and prompting a range of further questions about the primary subject matter. By asking us to see people as if they had certain inner states interacting in various ways, folk psychology offers an enormously powerful framing effect for explaining and predicting their behaviour (Toon, 2016; see also Demeter, 2013b). For example, if Barbara understands Dutch roll, we can infer that she has various beliefs about it. If we also assume that Barbara possesses various other mental states (e.g. the desire to avoid feeling sick), then we will also predict that she will act in certain ways (e.g. designing her new aircraft in ways that minimise Dutch roll). We can also ask whether, in order to possess understanding, the propositions that Barbara believes must be true (Elgin, 2009) or ask how exactly 'seeing' or 'grasping' a set of propositions differs from merely knowing them (Grimm, 2006).

The key advantage of fictionalism over eliminativism, then, is that it allows us to retain our ordinary categories for making sense of scientists' epistemic activities. Such talk serves its own distinctive purposes and need not await vindication from our final science of the mind. Of course, the fictionalist can still share the eliminativists' keen interest in the latest developments in cognitive science. If we want to understand exactly how it is that scientists recognise certain patterns, for example, or explore the way in which they interact with graphs or diagrams, the sentential framework may have little to tell us. Here we will need to consult our latest theories in these domains. And yet, the fictionalist insists, the sentential framework will continue to play its role, allowing us to draw together otherwise disparate forms of behaviour and explore their interconnections. Indeed, folk psychology might retain this role, even once our final cognitive science is in hand.

6 Conclusion

Eliminativism presents an important challenge to epistemologists and philosophers of science. Might the language that we use to describe reason, knowledge and understanding turn out to be fundamentally misguided? If so, must we find a radically new framework for describing our epistemic activities? Fictionalism offers a promising alternative. According to the fictionalist, our traditional, sentential framework plays a vital role in making sense of our epistemic practices, even if it fails to describe our inner machinery – even if, in fact, it was never intended to do that. For Churchland, the sentential framework is a bad theory. For the fictionalist, it is not a theory, but a metaphor – and an invaluable one.

Note

1 A noteworthy attempt to depart from the sentential framework is the semantic view of theories (e.g. Suppe, 1974; van Fraassen, 1980; Giere, 1988). For a critical discussion of the semantic view in relation to eliminativism, see Churchland (1989, pp. 157–158).

References

Beardsley, M. (1962). 'The Metaphorical Twist'. *Philosophy and Phenomenological Research*, 22 (3), 293–307.

Bechtel, W. (1996). 'What Should a Connectionist Philosophy of Science Look Like?', in R. McCauley (ed.) *The Churchlands and Their Critics*. Cambridge, MA: Blackwell, pp. 121–144.

Churchland, P. M. (1981). 'Eliminative Materialism and the Propositional Attitudes'. *The Journal of Philosophy*, 78 (2), 67–90.

Churchland, P. M. (1989). *A Neurocomputational Perspective: The Nature of Mind and the Structure of Science*. Cambridge, MA: MIT Press.

Churchland, P. M. (1996a). 'On the Nature of Explanation: William Lycan', in R. McCauley (ed.) *The Churchlands and Their Critics*. Cambridge, MA: Blackwell, pp. 257–64.

Churchland, P. M. (1996b). 'Bechtel on the Proper Form of a Connectionist Philosophy of Science', in R. McCauley (ed.) *The Churchlands and Their Critics*. Cambridge, MA: Blackwell, pp. 265–70.

Clark, A. (1989). *Microcognition: Philosophy, Cognitive Science, and Parallel Distributed Processing*. Cambridge, MA: MIT Press.

Clark, A. (2008). *Supersizing the Mind: Embodiment, Action, and Cognitive Extension*. Oxford: Oxford University Press.

Clark, A and Chalmers, D. (1998). 'The Extended Mind', *Analysis*, 58 (1), 7–19.

Demeter, T. (ed.) (2013a). 'Mental Fictionalism [Special Issue]'. *The Monist*, 96 (4), 483–638.

Demeter, T. (2013b). 'Mental Fictionalism: The Very Idea'. *The Monist*, 96 (4), 483–504.

Dennett, D. (1996). *Kinds of Minds*. New York: Basic Books.

Elgin, C. (2009). 'Is Understanding Factive?', in D. Pritchard, A. Millar and A. Haddock (eds.) *Epistemic Value*. Oxford: Oxford University Press, pp. 322–9.

Fodor, J. (1975). *The Language of Thought*. Cambridge, MA: Harvard University Press.

van Fraassen, B. C. (1980). *The Scientific Image*. Oxford: Oxford University Press.

Giere, R. (1988). *Explaining Science*. Chicago, IL: University of Chicago Press.

Giere, R. (2002). 'Scientific Cognition as Distributed Cognition', in P. Carruthers, S. Stitch and M. Siegal (eds.) *The Cognitive Basis of Science*. Cambridge, MA: Cambridge University Press, pp. 285–99.

Giere, R. (2006). *Scientific Perspectivism*. Chicago, IL: University of Chicago Press.

Grimm, S. (2006). 'Is Understanding a Species of Knowledge?' *British Journal for the Philosophy of Science*, 57 (3) 515–35.

Hempel, C. (1965). *Aspects of Scientific Explanation and Other Essays*. New York: Free Press.

Kvanvig, J. (2003). *The Value of Knowledge and the Pursuit of Understanding*. Cambridge, UK: Cambridge University Press.

Lycan, W. (1996). 'Paul Churchland's PDP Approach to Explanation', in R. McCauley (ed.) *The Churchlands and Their Critics*. Cambridge, MA: Blackwell, 104–20.

Menary, R. (2007). *Cognitive Integration: Mind and Cognition Unbounded*. Basingstoke: Palgrave Macmillan.

Menary, R. (ed.) (2010). *The Extended Mind*. Cambridge, MA: MIT Press.

Moran, R. (1989). 'Seeing and Believing: Metaphor, Image, and Force', *Critical Inquiry*, 16 (1), 87–112.

Nersessian, N. J. (2005). 'Interpreting Scientific and Engineering Practices: Integrating the Cognitive, Social, and Cultural Dimensions', in M. Gorman, R. Tweney, D. Gooding and A. Kincannon (eds.) *Scientific and Technological Thinking*. Mahwah, NJ: Erlbaum, pp. 17–56.

de Regt, H., Leonelli, S. and Eigner, K. (eds.) (2009). *Scientific Understanding: Philosophical Perspectives*. Pittsburgh: Pittsburgh University Press.

Riggs, W. (2003). 'Understanding "Virtue" and the Virtue of Understanding', in M. DePaul and L. Zagzebski (eds.) *Intellectual Virtue: Perspectives from Ethics and Epistemology*. Oxford: Oxford University Press, pp. 203–26.

Robbins, P. and Aydede, M. (2009). *The Cambridge Handbook of Situated Cognition*. Cambridge: Cambridge University Press.

Rowlands, M. (1999). *The Body in Mind: Understanding Cognitive Processes*. Cambridge, MA: Cambridge University Press.

Rumelhart, D., Smolensky, P., McClelland, J. and Hinton, G. (1986). 'Schemata and Sequential Thought Processes in PDP Models', in Rumelhart *et al.* (eds.) *Parallel Distributed Processing*. Cambridge, MA: MIT Press.

Suppe, F. (ed.) (1974). *The Structure of Scientific Theories*. Chicago, IL: University of Chicago Press.

Toon, A. (2014). 'Empiricism for Cyborgs'. *Philosophical Issues*, 24 (1), 409–25.

Toon, A. (2015). 'Where Is the Understanding?' *Synthese*, 192 (12), 3859–75.

Toon, A. (2016). 'Fictionalism and the Folk', *The Monist*, 99, 280–95.

Wallace, M. (2007). 'Mental Fictionalism'. Unpublished manuscript.

Walton, K. (1990). *Mimesis as Make-Believe: On the Foundations of the Representational Arts*. Harvard, MA: Harvard University Press.

Walton, K. (1993). 'Metaphor and Prop Oriented Make-Believe'. *European Journal of Philosophy*, 1 (1), 39–56.

Wheeler, M. (2005). *Reconstructing the Cognitive World*. Cambridge, MA: MIT Press.

Wilson, R. A. (2004). *Boundaries of the Mind: The Individual in the Fragile Sciences: Cognition*. Cambridge, UK: Cambridge University Press.

Yablo, S. (1998). 'Does Ontology rest on a Mistake?' *Proceedings of the Aristotelian Society, Supplementary Volume*, 72, 229–261.

10 Art, science and abstract artefacts

Steven French

1 Introduction

What is a scientific theory? It would seem that we can point to (figuratively and literally) any number: the Boveri-Sutton chromosome theory, Ragnar Nurske's balanced growth theory of economics, the molecular orbital theory of molecular structure and of course those old chestnuts, Maxwell's Theory of Electromagnetism, the Special Theory of Relativity, Quantum Mechanics ... the list goes on. But can we characterise or otherwise pin down what a theory *is* in general terms that go beyond simply listing examples?

A traditional answer insists that a theory consists of:

(i) an abstract formalism F;
(ii) a set of theoretical postulates (taken to be axioms) T;
(iii) a set of 'correspondence rules' C.

F incorporates a language L in terms of which the theory is formulated and a deductive calculus is defined. L contains logical and non-logical terms, where the latter are typically taken to be divided into observational and theoretical; the 'correspondence rules' then function as a kind of dictionary or bridge by relating the former to the latter. A 'partial interpretation' of the theoretical terms and the sentences of L containing them is then provided by the theoretical postulates – which contain only theoretical terms – together with the correspondence rules, which correlate the non-logical, theoretical terms with observable phenomena by allowing for the derivation of certain sentences containing observation terms from certain sentences containing theoretical ones. The partiality of the interpretation arises because the theoretical terms are not explicitly defined. This gives a certain degree of structural leeway that allows for the addition of further correspondence rules as science advances, thus extending the interpretation of these terms.

The deficiencies of such a view are well known – the reliance on correspondence rules, the dependence on language and, perhaps more generally, the generation of numerous problems that have to do with the nature of the formal framework itself, rather than as arising out of scientific practice. As a result, this view, once so entrenched as to be called 'the Received View' (Suppe, 1989), has long since been abandoned (although for signs of a revival, see Lutz, 2014). The 'new orthodoxy', as it is called by advocates and critics alike, is the so-called 'Semantic Approach', at the core of which is the claim that '... theories are not collections of propositions or statements, but rather are extra-linguistic entities which may be described or characterised by a number of different linguistic formulations' (Suppe, 1977, p. 221). Typically, this core claim is cashed out by taking theories to be classes of set-theoretical models characterised by what their linguistic formulations refer to when the latter are interpreted semantically. It is in this sense that theories may then be understood as extra-linguistic, and in turn it may then be claimed that, according to this approach, theories *are* nothing more than families of such set-theoretic models. And, as with the Received View, certain devices can be introduced to accommodate the openness and partiality of theories in scientific practice (see da Costa and French, 2003).

However, by identifying theories in terms of classes of models, the Semantic Approach just pushes the answer to our question – what is a theory? – back one step (as does the Received View, of course). Still, that identification does make the answer straightforward: a theory is whatever these models are, and as set-theoretic, they are generally taken to be abstract entities.

2 Models as abstract entities

Perhaps the most well-known defender of this view is Giere (see, for example, 1988, p. 80ff). According to him, models are generated from certain principles plus specific conditions, and the attempt to apply them to the world yields hypotheses about their 'fit' to systems in the world. By 'principles' here, Giere means what are typically understood to be empirical laws, such as Newton's laws, Maxwell's laws of electromagnetism or the laws and principles of Special Relativity. However, he insists, these should not be regarded as vehicles for making empirical claims, given their considerable generality. If we were to insist on taking them to be genuine statements about something, we would have to say what it is they are about, and he argues that the only candidates are 'highly abstract objects' in the sense of '... an object that, by definition, exhibits all and only the characteristics specified in the principles' (ibid., p. 745). The principles are then true, but only trivially so, of such objects and act as templates, with the aid of specific conditions, for the construction of more specific abstract objects, namely, models. To get from these to actual empirical claims we must then 'designate', or denote, a particular concrete system, as when we apply the model for simple harmonic motion to an actual pendulum, for example.

In what sense are scientific models abstract? Giere answers as follows[1]:

> First, they are abstract objects like numerical relationships or geometrical figures, square roots, perfect squares and circles, or never constructed buildings described in architect's drawings. They are not physically realized.

> Second, they are abstract in that they are not fully specified. Newton's Laws refer to forces, masses, accelerations, velocities, positions, and times, but not to any specific such objects or quantities.
>
> (Giere, 2008, p. 5)

The second sense of abstraction just has to do with their place in the hierarchy between data models and theories, and the notion of abstraction, although much debated within philosophy of science and usually in the same context as that concerning idealisations (ibid.), does not directly pertain to the models' ontological status. The first sense, however, rests on a comparison with other abstract entities, albeit of different kinds, and obvious questions arise: if models are like mathematical entities, do they too exist in some Platonic realm? And how, then, are they accessed? In particular, by what means would scientific theories, *qua* abstract objects, be *discovered*? Presumably a proponent of such a view would have to appeal to some kind of 'platonic' sense, similar to that by which mathematicians supposedly discover their theorems on the Platonists' account. But even if that is plausible for mathematical objects, it seems much less so for scientific theories, whose heuristic development can be much more straightforwardly traced. Certainly we do not find in science many examples of the kind of first-person evidence regarding this means of discovery that appear to drive certain Platonist views of mathematics.[2]

One possibility would be to argue that access to scientific theories as abstract objects could be understood as mediated via these heuristic moves by which they are 'discovered'. However, such 'moves' differ considerably in nature and kind. Consider Post's famous taxonomy, for example, which includes very general criteria, such as the General Correspondence Principle, according to which parts of the structure of new theories should correspond to the empirically well-supported parts of their predecessors, as well as more specific suggestions to do with the role of symmetry principles in physics, etc. (Post, 1971). It is difficult to see what these have in common, both between themselves and with the sensory modalities in terms of which we gain access to observable objects, which is the analogy that Platonist accounts of 'discovery' typically appeal to. Of course, one might argue that the relevant comparison should be with our indirect access to unobservable entities, but again it isn't clear that there are any similarities with the inferential moves that are made there. So, until this mode of access to theories is spelled out in terms of the specific heuristics, this proposal remains just a suggestion.

But Giere also suggests that models are akin to the 'never constructed buildings described in architectural drawings' (Giere, 2008, p. 5), which raises another set of questions: what is the nature of the relationship between that drawing and the relevant abstract entity? Does the act of making the drawing create the entity? What happens to the abstract entity when the drawing is destroyed? Similar questions can be posed for models and, for example, equations in a scientific paper or, more worryingly perhaps, on a whiteboard. It is these questions I shall focus on here and I shall consider various options for answering them, drawing on both the history of the philosophy of science and the philosophy of art. In the end I shall conclude that these answers themselves raise further questions that collectively cast doubt on the plausibility of the claim that theories and models are abstract entities.

3 The surprising life of a World 3 entity

As is well-known, but these days seldom discussed, Popper argued for the reality of a world over and above that of physical things, which he designated 'World 1', and also beyond that of 'World 2', the world of mental processes. His 'World 3' was the world of the products of the human mind and was occupied by scientific theories as well as works of art (including, notably, musical works) and social institutions (Popper, 1978). Thus, he writes, 'Examples of World 3 objects are: the American Constitution; or Shakespeare's *The Tempest;* or his *Hamlet;* or Beethoven's *Fifth Symphony*; or Newton's theory of gravitation (ibid., p. 145).

Not surprisingly, the relationship between objects in Worlds 3 and 1 depends on the nature of the object concerned.[3] Thus, World 3 accommodates sculptures such as Michelangelo's *The Dying Slave* (http://en.wikipedia.org/wiki/Dying_Slave), that exists both in World 1 as a block of marble and in World 3, as the creation of Michelangelo's mind (Popper, op. cit., p. 144). Likewise, a painting such as *Rain, Steam and Speed* exists as a painted piece of canvas in World 1 and as the result of Turner's imagination in World 3.

Here the first of our questions arises: what happens when an artwork is destroyed, such that it can no longer be said to exist in World 1? Consider Michelangelo's early piece, *Sleeping Cupid* (https://en.wikipedia.org/wiki/Cupid_(Michelangelo)#Sleeping_Cupid), created as part of a scam and subsequently presumed destroyed in the great fire of the Palace of Whitehall in 1698: does the artwork nevertheless still exist in World 3? It would defeat the point of Popper's distinction if *Sleeping Cupid*, as a World 3 entity and thus a product of Michelangelo's mind, ceased to exist when Michelangelo did. But does it still exist there once its World 1 counterpart has gone? Of course, we might try and recreate it if, for example, we were to discover a sketch of the piece (see: http://www.michelangelo-gallery.com/michelangelo-drawings-list.aspx). But then *that* statue would not be the creation of Michelangelo's mind, and a different World 3 object would be created, however similar to the 'original' (or so it would seem; we shall come back to this issue). To maintain that artworks persist in World 3, no matter the vicissitudes of World 1, is to grant them more of a Platonic nature than seems compatible with Popper's view of theories, for example, but to deny this raises certain obvious issues that I shall return to below.

When it comes to 'repeatable' works of art, such as literature, the relationship is more akin to that between types and tokens: there are many different copies of *The Tempest*, scattered about World 1, but insofar as they contain the same text, they are all the embodiment or physical realisation of 'one and the same book' that 'lives' in World 3. One can then extend this more or less straightforwardly to theories – think of Newton's *Principia*, for example, with its multiple copies in different libraries. Significantly, given what we just considered with regard to Giere's stance, Popper draws this distinction as follows: 'One can, if one wishes, say that the World 3 objects themselves are *abstract* objects, and that their physical embodiments or realisations are *concrete* objects' (ibid.).

However, the relationship between these two kinds of objects is not to be conceived of in Platonistic terms. First of all, the objects of World 3 are not timeless; indeed, Popper takes it as crucial for his overall epistemology that theories, *qua* World 3 objects, be seen as tentative and hence as subject to change. He writes,

> In opposition to Plato and Hegel I consider *tentative theories* about the world—that is, hypotheses together with their logical consequences—as the most important citizens of the world of ideas; and I do not think (as Plato did) that their strangely non-temporal character makes them eternal and thereby *more real* than things that are generated and are subject to change, and to decay. On the contrary, a thing that can change and perish should for this very reason be accepted as prima facie real; and even an illusion is, *qua* illusion, a real illusion.
>
> (Popper, 1972, p. 300)

Second, these objects interact causally with those of the other worlds: '[T]hey [the objects of 'World 3'] may be real in that they may *have a causal effect* upon us, upon our world 2 experiences, and further upon our world 1 brains, and thus upon material bodies (1978, p. 150).

Clearly, then, World 3 objects are different from mathematical ones, as conceived in Platonistic terms, since they can be created and are not causally inert.

It is this causal efficacy of theories that underpins Popper's 'fundamental argument' for his World 3 ontology that invokes the way in which science has changed the (physical) world, or, more particularly, the way in which 'conjectures and theories' are used as 'instruments of change':

> [S]cientific conjectures or theories can exert a causal or an instrumental effect upon physical things; far more so than, say, screwdrivers or scissors.
>
> (ibid., p. 154)

Given his causal criterion of reality, namely that '… what is real or what exists is whatever may, directly or indirectly, have *a causal effect* upon physical things …' (ibid., p. 153), such conjectures and theories must be taken to be real. And if they are real they must exist somewhere. But they can't exist in World 1, for obvious reasons, nor can they be placed in World 3, since they can be objectively evaluated – just as performances of Beethoven's *Fifth* can be judged to be 'good' or 'bad', so theories can be determined to be true or false, empirically adequate or not.

Furthermore, theories cannot be mental objects because they include their consequences, but some of these consequences do not exist in the minds of the 'creator' (Popper's term) of the theory nor, indeed, do they exist in any mind:

> There are many important logical consequences of the Special Theory of Relativity which Einstein did not think of in 1905; and there may be important logical consequences of this theory which nobody has thought of so far, and which perhaps nobody will ever think out.
>
> (ibid., p. 158)

Indeed, if to fully understand a theory is to grasp all of its logical consequences, and given that these will be infinite in number, nobody can have such a complete understanding, '... which shows again that the theory, in its logical sense, is something objective and something objectively existing – an object that we can study, something that we try to grasp' (1972, p. 299).

Finally, and relatedly, Popper argues, theories have a property that *only* existing things could have: the element of *surprise*. Thus, he writes,

> Such a theory, or such a system, is infinite, and may be full of surprises. Thus it must have been a surprise for Einstein when he found, shortly after writing his first paper on Special Relativity, that the now-famous formula $E = mc^2$ could be deduced from it as a theorem.
>
> (1978, p. 162)

This aspect of 'surprise' is taken as a mark of the reality of something: just as physical objects surprise us as we discover more about them, so too do scientific theories.[4]

Drawing on this episode from the history of the philosophy of science is helpful in articulating the sense in which theories might be regarded as 'abstract' entities, but as we have seen there are still questions that need to be answered. Given that Popper includes artworks in his World 3, it seems an obvious move to turn to the philosophy of art in order to see if there are any moves or devices that we can appropriate to help make further sense of this approach. With that in mind, let us now consider that view of artworks that takes them to be 'abstract artefacts'.

4 Models and theories as abstract artefacts

As we have seen, one fundamental question that has arisen is the following: what is the relationship between the 'creator', whether of the artwork or the theory, and the abstract entity that is created? Thomasson has responded to this challenge by characterising certain artistic objects, such as works of music and literature, as 'abstract *artifacts*' in the sense that although they lack a spatio-temporal location, they are still created, come into existence, change and may cease to exist (Thomasson, 1999; Thomasson, 2004; other works such as painting and sculptures, obviously, count as concrete artifacts). It is through their creative activities that artists bring such artworks into existence, and their continued existence depends on the artist's intentionality. As Thomasson says, this undermines the traditional bifurcation into abstracta and physical entities, and it offers what might appear to be a useful metaphysical suit in which to clothe the views of theories adopted by Popper, and, more recently, Giere.

However, there are obvious concerns that need to be dealt with in appropriating such an account.

First of all, one of the prime motivating features behind this approach is that works of literature and music are brought into existence at or by a particular moment in time via creative activity. Of course, it is not being suggested that such coming into being is instantaneous or even takes place over a very short space of time. Although Mozart was famous for being able to improvise 'on the spot', there is apparently little or no evidence that these improvisations led to more well-known and established compositions (see: https://en.wikipedia.org/wiki/Mozart%27s_compositional_method).[5] With regard to the latter, he apparently took two days to write his twenty-fifth symphony, while Beethoven spent six years on his Ninth. *The Lord of the Rings*, famously, took many years to complete, with Tolkien starting to write the novel in 1937 and it eventually being published in 1954/55. The point, of course, is that there comes a point (!) when the composition is deemed to be finished and the artwork can be said to have been brought into existence.[6] Should this also be considered a fundamental requirement on the ontology of theories?

Here one might balk when faced with the subsidiary question: did General Relativity just pop into existence when Einstein thought it up? Note that no one is suggesting – at least not here – that curved space-time popped into existence when Einstein came up with GR! What, then, about the 'laws' of GR, embodied in the famous tensor equations? Here things get a little (but only a little) more nuanced: if you are a realist about laws and take them to be modally informed features of the world that 'govern' the phenomena, in some sense, let me reassure you that it is not being suggested that these likewise 'pop' into reality through Einstein's creative endeavour. If you are a Humean, then you might be inclined to think they do pop into reality, inasmuch as the theory could be said to do so, since laws or, better, law statements are just shorthand descriptions of repeatable phenomena (which, again, no one is suggesting are doing any kind of popping).

With that out of the way, the balking might subside – we are talking about whether the theory as an *entity in itself* could be said to come into being through Einstein's creative acts. And on the face of it, there seems to be a reasonable degree of similarity between artworks and theories in this regard. So, just as musical compositions can be said to be created over a period of time, we can think of Einstein's theory as being developed from 1907, but finally coming into existence in 1915, with his presentation of the field equations to the Prussian Academy of Sciences. And, heading back in the other direction, just as we can trace the development of the theory through Einstein's notes, papers, letters, etc., thereby extracting the heuristic moves he took, so we can do something similar with Mozart or Beethoven.

But how does engaging in a certain practice – artistic or scientific – or following a certain heuristic move *create* an artifact that is abstract? In the case of a concrete artifact, such as a painting or a physical model, one can answer this question, not least because the practices and moves considered will embody certain causal relationships that one can understand as affecting the final product, namely the painting or model. One can see how Picasso's studies of the gored horses in the bullring relate to the horse in agony in *Guernica*, where that relationship is effectively mediated by the physical movement of his hand with the brush against the canvas and so on. Likewise, one can track the various moves that led Crick and Watson to construct their tinplate and wire model of DNA (see, for example, Schindler, 2008), where again these moves are translated into their positioning the tinplate and wire in a certain way, at a certain angle, of a certain length and so on. But how is that effectivity established in the case of abstract artifacts? Here we face well-known problems, and for all that it might be insisted that we must not beg the question against this position and that a new, third category of entity is being articulated, we are surely owed some account of how we can bridge that final gap, between the practice and the entity.

One way to do that might be to say that insofar as the creation of such a theory involves the *intention* of the scientist, it is this intention alone that brings the theory into existence. Again, however, one might wonder how this meshes with the heuristics of scientific discovery. At the very least, anyone adopting such a view would have to acknowledge that these intentions would have to be constrained in some way, on pain of allowing massive overpopulation of the realm of such artifacts. Or perhaps one could argue that any intention to produce a theory, no matter how bizarre, unjustified or out of step with current science, creates the corresponding abstract artefact, but that only the sub-set of the resultant plethora of such artefacts that meet the relevant heuristic criteria would count as 'theories'. And again one would have to accept certain relations between the various artefacts corresponding to the different stages of the development of the theory, paralleling the relations we discern between their material counterparts in practice.

A closely related option would be to argue that although creation *does* require causation (in some sense), and we can't causally interact with abstract objects, this in itself doesn't mean that we can't create abstract objects. After all, during the process of creation the abstract object doesn't yet exist, so one is free to tell a story where we manipulate things we *can* causally interact with, and the abstract object comes about as a result of our manipulating these things.[7] But this just pushes the bump in the rug around, and again the question arises: what is this 'coming about'?

Of course, one could simply deny that creation requires causation. Thus, it might be claimed that at the same instant that a scientist 'discovers' or creates a theory, the corresponding abstract object simply comes into being. One might then speculate whether it comes into being at the moment the scientist conceives of the theory in her mind, or when she writes down the relevant statements, equations, etc. or whether it 'emerges' during such a process. This broadly meshes with what can be called the 'lightbulb' or 'Eureka!' view of scientific discovery, which features quite prominently in scientists' own autobiographical reflections. Such a view, insofar as one can say anything philosophically interesting about it, tends to overlook the kinds of heuristic factors that have been well-documented in the history of science (again, see Post, 1971). However, one could allow for such factors by modifying the above line so that in conceiving of a hypothesis, say, or writing down an equation, etc. on the basis of making the relevant heuristic moves, one thereby creates the corresponding abstract element. Such modification raises familiar problems: does the abstract artifact only come into being at the end of the heuristic process? Can we pin that down? In Einstein's case, was it really when he presented the *General Theory of Relativity* to the Prussian Academy of Sciences? That seems somewhat arbitrary, as if the theory could not be considered to be in existence when he carried his notes or paper into the lecture theatre, and it popped into being when he presented it (again, when he began the presentation or when he finished it?). Or did it come into being when he wrote the final sentence, or the final full stop? Or when he completed it 'in his head'?! [8]

And what about all the steps leading up to that final moment? Suppose some steps led to something akin to the final theory, but it was not quite there yet. We know that in Einstein's case the infamous hole argument[9] led him to abandon his focus on general covariance, even publishing a relativistic theory of gravitation that was not covariant in 1913, only for him to return to this requirement when he took the argument (erroneously) to be a trivial error (see Norton, 1984). More generally, we can find other examples in the history of science of 'proto-' theories, of not just speculations but theories that are almost in their final form, but aren't quite 'there' yet. Are we to say that all of these 'almost rans' are also created in Thomasson's realm or Popper's World 3 as a result of the relevant practices or heuristic moves being enacted? And even if we could find a principled way of excluding them, we would surely have to acknowledge that some combination of such elements in practice can be taken to compose the theory concerned. And so, paralleling this, at the abstract level, we would have abstract objects corresponding to such elements composing the abstract object corresponding to the theory.

Two further worries arise as a result. The first is that the relevant practices are just too vague and diverse to support the coming into being of an entity in any clear or well-defined manner. The questions begin to proliferate: which practices? Over what time frame? How are their entwined interrelationships reflected ontologically in the artifacts created? And so on. The second has to do with the ontological inflation involved as our World 3 becomes cluttered with not just discarded theories but also those that didn't quite make it, like Einstein's of 1913 as well as those that developed into those that 'made it', the blind alleys and false starts, the detours and digressions .... Indeed, unless some sort of exclusionary principle is applied the realm of abstract artifacts is going to be chock-full of entities, but it is hard to see what might justify such a principle.

Perhaps these worries are not so severe when it comes to artworks (how many drafts of *Hamlet* did Shakespeare write?) but they give us pause in thinking of scientific theories as abstract artefacts in this way. And that pause may be deepened by a further feature that could be seen as distinguishing theories from artworks. As we shall see, the gap that this opens between them can be closed by denying the existence of this feature, but then reflection upon the underlying motivating intuition prises it open again.

5 Multiple discovery and modal flexibility

This feature has to do with 'multiple discovery', a phenomenon that, in the history of science, at least, has apparently been repeatedly (and thus multiply) discovered by historians and sociologists of science, as Merton famously noted (Merton, 1963).[10]

Here are just three examples, taken from different sciences: the *Special Theory of Relativity*, the *Theory of Evolution* and the *Theory of the Asymmetric Carbon Atom*. The first, or so it has been claimed, was co-discovered by both Einstein and Poincaré (Zahar, 2001), the second by Darwin and Wallace (van Wyhe, 2013) and the final example, by van 't Hoff and LeBel (Gay, 1978). One can give various explanations for this 'phenomenon': a sociologist might point to the relevant historico-cultural context as a kind of common cause of such multiple discoveries (or at least those that are, or nearly are, simultaneous), whereas a philosopher of science might seize on certain 'internal' factors having to do with the initial choice of problem, the limited range of heuristics available and, more generally, the deployment within a given field of a more or less agreed (if only tacitly) methodology (involving both heuristics and justification) over a period of time. And, going further, a realist might argue that given the way the world is, we should not be surprised that more than one scientist comes up with the same or similar ways of describing it.[11]

Now, at first blush, however we explain the 'fact' of multiple discoveries, it seems to raise further problems for the suggestion that the creation of theories as abstract artifacts is mediated by the intentions of the scientists concerned. It hardly seems plausible that the intentions of Poincaré and Einstein, or Darwin and Wallace, van 't Hoff and LeBel, could be the same as one another and thus, even if we could dispel the above mystery, we must account for how different intentions could produce the same theory (do the intentions become entwined in some way?!). Even if we downplay the role of intentions and fall back on practices, as urged by sociologists, we're not out of the woods. Although it might seem plausible that the historical context and associated practices that Poincaré and Einstein were engaged in were sufficiently similar as to lead, somehow, to the creation of the same theory, it stretches the point to claim that the same holds for Darwin and Wallace, given their different social statuses, cultural context, geographical travels and so on. As for van 't Hoff and LeBel, both the rationales and inspirations were different: the former was primarily concerned with isomerism and was following a line established by Kekule, whereas the latter set himself the problem of explaining optical activity and followed Pasteur (Gay, op. cit., p. 222). Of course, one might try to give a more sophisticated account of the relevant practices that would reveal what they had in common. Thus, there may be methodological commonalities (ibid.) that one could appeal to, but insofar as these are intended to cover other theories as well, they cannot be appealed to in order to account for the creation of these ones in particular.

There does not seem to be a similar phenomenon in the history of art; that is, we do not find, it seems, that as well as Picasso, another artist created *Guernica*, or a painting indistinguishable from the latter, or that someone other than Melville wrote a work identical in content to *Moby Dick* or that as well as Beethoven, someone else also composed the *Ninth Symphony*. Of course, there are precursors or works that deal with similar themes, or yield similar insights or whatever. Consider, for example, the account by Jeremiah Reynolds of an aggressive (justifiably) white whale in his 'Mocha Dick: Or The White Whale of the Pacific: A Leaf from a Manuscript Journal' (https://en.wikipedia.org/wiki/Mocha_Dick); for all that this influenced Melville, it is not, of course, the same literary work as *Moby Dick*. And it should not come as much of a surprise that we do not typically find multiple creations in the history of literature. If we take as part of the relevant identity conditions the structure of the work, or most restrictively, perhaps, the word sequence[12], then it is clearly implausible to expect that, even over a reasonable length of time, a book might be created that is identical to *Moby Dick*, *Ulysses* or *Don Quixote* (and similar considerations can obviously be applied to musical works).[13] But, of course, we do not find those conditions applied to scientific theories – it is not the case that every presentation of the Special Theory of Relativity, for example, must reproduce the word sequence or even the structure of Einstein's 1905 paper!

Granted that flexibility, however, multiple discovery does not quite draw as clear a distinction between artworks and theories as might first be thought. Indeed, many, if not most, historians and philosophers of physics would now agree that whatever it was that Poincaré came up with – theory, proto-theory, hypothesis – it was *not* the Special Theory of Relativity. Thus, Poincaré and Einstein had different attitudes towards absolute motion – the former taking it to be undetectable, the latter as non-existent – and the Lorentz transformations – the former taking them to be merely calculational devices, defined only with respect to the ether rest frame, the latter as representing physical relationships between coordinate systems (for an overview see Adlam, 2011). More importantly, perhaps, Poincaré did not see the relativity principle as explanatory in and of itself, but rather as an empirical summary that could be used to confirm or refute various hypotheses (ibid.; Katzir, 2005). Einstein, on the other hand, took the principle to be a fundamental constraint on the form of the relevant laws, and this represents a fundamental shift in what was taken to require explanation in this context, akin to that which marks the difference between the Aristotelian and Newtonian approaches to what we now call inertial motion. And there is the further related difference regarding the associated ontology, with Poincaré retaining the electromagnetic ether and Einstein dispensing with it, presenting the theory initially – and famously – positivistically, in terms of rods and clocks, before reluctantly accepting Minkowski's spacetime formulation.[14]

Likewise, although Darwin and Wallace themselves thought their theories were identical, various historians of science have noted crucial differences. Thus, Darwin's theory is concerned with *competitive* selection between members of the same species, such that the less 'fit' lose out to the 'fitter', whereas Wallace's theory is about *environmental* selection, whereby an individual has to survive the rigours of a particular environment (Nicholson, 1960); the former is about competition between individuals and the latter competition between varieties (Bowler, 1976); according to Darwin's theory, an advantageous variation would increase and its parental form decrease in frequency until it became extinct, whereas according to Wallace's, both would exist, albeit with the former as more numerous than the latter, until some environmental change forced the extinction of the parental form (Bulmer, 2005). Again, we don't need to get into the details or the dispute between historians of science; what is clear is that an argument can be made that the theories are different despite similar appearances and, indeed, the views of the discoverers themselves (ibid., p. 134).

Thus, one might suspect that the claims about multiple discovery are overplayed. Once the irrelevant cases are stripped away, such as the multiple invention of blast furnaces at one end of the spectrum and the co-discovery of the calculus at the other,[15] there are precious few, if any, examples left standing. But in that case theories start to appear much closer on the spectrum to artworks again, with the role of the intentions of the relevant scientist (Einstein's being different from Poincaré's, Darwin's from Wallace's and so on) akin to those of the author or composer, at least when it comes to their relationship with the artefact concerned.

But perhaps the gap can be widened once again if we think further about the intuition that underlies the plausibility of multiple discovery, namely that 'someone else could have done that'! Suppose we were to ask the following questions: Could Braque have painted *Guernica*? Or Tolkien written *The Chronicles of Narnia*? Could Poincaré have discovered Special Relativity? Intuitively we might be inclined to answer 'yes' to all three – an intuition motivated perhaps by the artistic closeness of Braque to Picasso, at least initially, of the physical and literary proximity of Tolkien and Lewis and of the scientific relatedness of Poincaré and Einstein. Take the first two: It would seem that we can more or less easily imagine such scenarios: we can imagine Braque having been commissioned to produce a mural for the Exposition Internationale des Arts et Techniques dans la Vie Moderne (although perhaps not by the Spanish Republican Government) and reading the horrific eye-witness reports of the bombing in the Times or New York Times. We can likewise see, in our mind's eye, Tolkien seated in the Eagle and Child pub, reading passages from one of the Narnia books to Lewis and his friends.

But of course imaginability, easy or otherwise – or, more generally, conceivability – does not imply 'genuine' possibility. There is an issue as to how naturalistic we should let our modal metaphysics be, but for our purposes, as soon as we start to flesh out some of the details of these 'imaginings' their plausibility fades (Rohrbaugh, 2005, p. 215). Would Braque have had the detailed knowledge of bullfighting to be able to produce those anguished animal figures? Or, if we chose some lesser known Spanish painter, would he or she have had the skill and relevant artistic background?[16] There is more to say, of course (see Rohrbaugh, ibid.), and some tricky examples to tackle, but in general artworks do not seem to be 'modally flexible' in the sense that they could have been produced in a different way or by a different artist and yet still be *the same* artwork.

The issue now is whether theories are more flexible in this regard. Consider the third question above: one could maintain that if Poincaré had been in the same context, scientific and otherwise, as Einstein, and, crucially, had changed his attitudes in certain ways, then it is plausible to imagine that the theory he would then have come up with would have been identical to Einstein's. But that seems just the same as insisting that if Braque had attended a lot of bullfights and had developed the same techniques as Picasso and had acquired other artistic attributes, then he could have created *Guernica*. Perhaps one might be inclined to argue that in the case of Braque and Picasso, the former would have had to have had the entire life experiences of the latter in order to create *Guernica* – in effect, he would have had to *be* Picasso – but that is not the case when it comes to Poincaré and Einstein; Poincaré could have remained French, could still have proved the Euler-Poincaré theorem and yet could also have discovered Special Relativity. Thus, Poincaré could have had the intention to produce Special Relativity without having been Einstein, but Braque could not have had the intention that created *Guernica* as an abstract artefact without having been Picasso. That might seem a slim basis on which to hang a difference between theories and artworks, but we could bolster it by insisting that even allowing for those factors that relate to the actual differences between what Poincaré came up with and Special Relativity, the latter can accommodate a broader range of modal factors pertaining to its 'production' or discovery than most artworks. And ultimately, of course, this is because the theory, unlike the artwork, in latching onto reality (in whatever sense that is explicated) is determined by something (reality, the world, whatever) that goes beyond the intentions of Einstein, Poincaré or whoever, and thus could have been arrived at by someone other that Einstein. Of course, this is to betray a certain realist predilection (as touched on above), but even if one were an anti-realist of some kind, the requirement of empirical adequacy alone could be argued to be enough to raise doubts about the role of intentions here. Either way, one could argue that theories are, indeed, modally more flexible than artworks, in this sense. But then that seems to weaken the link, via the relevant intentions, with the creator/discoverer.

6 Persistence

And of course, it is not just that abstract artefacts are *brought into existence* via our intentions, but they are also *maintained* by them. Thus, the idea (as we noted with Popper) is that once created, such artefacts are not independent of us or exist eternally, but, rather, their ongoing existence depends on some relevant intentions being maintained. Now it is not entirely clear what that last phrase amounts to. Consider a piece of music, such as one of the 'lost' compositions of the young Mozart (see: http://www.bbc.co.uk/news/entertainment-arts-17493329). Suppose, first of all, that all it takes for one of these compositions to be brought into existence as an abstract artefact is for Mozart, having composed it 'in his head' to have the relevant intention to play it, again 'in his head' (indeed, he may have mentally played it through, hearing all the sounds on his piano). Is it the case that the piece of music is then maintained in existence by Mozart thinking about it? That would surely render it too fleeting and ephemeral! But suppose it remains in existence as long as Mozart can recall it, or bring it up from memory – presumably that means it ceases to exist when Mozart permanently forgets it, perhaps through illness, and certainly, it would seem, when he dies.

Now suppose that what it takes for the music to be brought into existence as an abstract artefact is not just that Mozart composes it in his head, but that he also writes it down as a score. One might then insist that the artwork persists in existing as long as the score exists, since the latter provides the means of producing a performance of the music. One might further extend this from the score to a recording, or just a memory (again, doubts might arise) or some combination of these. But questions as to the relationship between the score, the performance, the relevant intentions and the artefact intrude once again. Suppose the notebook containing Mozart's childhood composition had not been found and, indeed, was never found. Would the artefact persist in existing even though the composition was never played? In which case, what is the role of the score in this relationship? Suppose the score gently moulders away into dust – does the artefact likewise fade into the dusk in World 3?

Or suppose that what it takes is not just that the music has to be composed and written down or noted in some way, but it also has to be performed in order to sustain the existence of the artefact. Now the question is, does it just take one performance to sustain it? And for how long? Does it cease to exist after a certain period (surely not)? Or when the means to perform it no longer exist and/or can never be re-created (for example, following the worldwide banning of all pianos or the means for creating them!). Or perhaps the piece as an abstract artefact is brought back into existence each time it is played? Again, ephemerality, not to mention absurdity, beckons.

Similar questions arise when this view is applied to theories and models, but, again, there are differences when compared to artworks. So, consider Einstein's 'lost' precursor to Hoyle's steady-state theory of the universe, written on notepaper during a trip to California in 1931 (Castelvecchi, 2014).[17] Again, we can ask what it was that sustained this theory as an abstract artefact – was it the mere thought? The writing down of the relevant equations? The equations plus an interpretation? Did it remain in existence all the time the manuscript was lost? Or did it pop back into existence when the manuscript was found? Or, more fine-grainedly, when not only the manuscript was found but when it was understood (after all, it could have been discovered by a musician)? Again, it is impossible to begin to answer these questions until we have a better idea of the nature of the purported relationship between the intentions, the theory and its concrete manifestation.

Here once again one might appeal to the different epistemic status of theories and artworks. In Einstein's case, the manuscript concerned was probably put aside and 'lost' because Einstein realised that he had made a mistake in his calculations and the theory simply was not empirically adequate (as revealed by his crossing out part of the relevant calculation; see Castelvecchi, ibid.). Now, one could insist that some musical works, for example, are likewise put aside because of internal incoherence, or because the melody just isn't strong enough or the overall theme just doesn't work, but empirical adequacy clearly seems a different sort of factor when it comes to the withdrawal of intentions.

This offers both an opportunity and further challenges. One might seize on this point and suggest that theories as artefacts go out of existence when they are abandoned by their authors, for any of the reasons indicated above. Unlike artworks, this might provide an appropriate 'end point' for the theory concerned. But this raises further concerns: does a theory have to be abandoned only by its author or by everybody to cease to exist? Or, conversely, what about cases where only the author clings to the theory in the face of opposition, only to be subsequently vindicated? Is that desperate belief enough to sustain its existence as an abstract artefact? Here we might think of Wegener's continental drift hypothesis, for example: although it is perhaps too crude to say that Wegener was alone in maintaining his belief in the theory, the majority of geophysicists rejected it through the 1940s and 1950s, particularly in the US (see Frankel, 1987). What about cases where a theory is generally abandoned, only to be rediscovered and revived? Does it come back into existence? Or does it cease to be when the flaw is discovered, whether by the author or someone else? In that case, could we have a situation where a theory as abstract artefact continues to exist for many years, perhaps forever, even though internally incoherent, say, as long as no one discovers that fact? And the blurred boundary between this kind of process and the domain of heuristics raises further issues, since not only may someone have the idea for a theory only to almost immediately discover that it is flawed, leading to a flickering existence at best of this view, but tackling a perceived flaw in a new theory is often part of the heuristic process[18], as the theory is honed and shaped, typically prior to publication or presentation. Again, do we have just one artefact that is shaped and refined during such a process or, in line with what is suggested above, does the theory vanish from World 3 when the flaw is discovered and a new, flawless artefact subsequently created?

This brings us on to a further point: unlike most musical works, or artworks in general, there may be questions as to where one theory ends, as it were, and another begins. Thus, consider Bohr's model of the atom and Sommerfeld's 'extension' of it via the quantisation of angular momentum, which rendered the electron orbits elliptical rather than circular, thereby allowing for quantum degeneracy (see Eckert, 2014). This is typically portrayed as the pinnacle of the 'old' quantum theory, before the new matrix and wave mechanics of Heisenberg and Schrödinger changed the landscape completely. Now, was this merely an adjustment to or at best an extension of Bohr's model, or did it constitute a new model in its own right? It is referred to in both ways, and there is no obvious criterion by which to determine whether it is one or the other. Likewise, how does post-Darwinian evolutionary theory relate to Darwin's work? Is it merely a development of the same theory, or something else?[19] Or perhaps we shouldn't be talking about theories and models in this context at all, but about research programmes or even, heaven forfend, paradigms and such.

Do we see anything similar in art? A useful example would be that of comics, which make a potentially interesting point of comparison with theories since, like novels, they are standardly (but not, perhaps, necessarily) 'multiple' works in the sense of being repeatable and admitting of instances rather than mere copies (Meskin, 2012: p. 32). However, unlike novels, comics are 'encoded' rather than 'exemplar based', in the sense that neither the original art nor the engravings or web code or whatever subsequently produced counts as an instance of the comic itself (ibid., p. 38).

We might think of a theory in this way, as 'encoded' in the abstract artefact, or whatever is in World 3, say, from which instances can be obtained via a similar sort of process, resulting in the theory's reproduction in a book, or a scientific paper or a Powerpoint presentation. And, like comics, theories are – or may be, at least – hybrid in the sense that they may include linguistic, mathematical and pictorial elements. But the important similarity here, of course, is that comics offer obvious and numerous examples of works that are initiated by one author but continued by others: so, consider a recent incarnation of the Marvel comic *Moon Knight*, initially written by Warren Ellis and drawn by Declan Shalvey, replaced by Brian Wood as writer with Greg Smallwood as artist and since replaced again. The character is the same, the series is the same, so the general framework, overall plot, issues tackled, hero methodology, etc. are all the same, but, of course, certain plotlines are further developed, new ones are introduced, new challenges are offered and so on. So, is the Wood/Shalvey *Moon Knight* the same or a different artefact from the Ellis/Smallwood one?

On the one hand, with a different writer, and given the interdependent relationship between writer and artist when it comes to the creation of comics, one might be inclined to argue that the change marks the creation of an entirely new artefact. On the other, one could argue equally well that the Wood/Shelvey collaboration represents a development of the comic as a series, similarly to the way the Sommerfeld model represents an extension of the Bohr system. And, as in the case of theories and models, when faced with a choice of line to take, one might feel both that the abstract artefact view is, at the very least, unhelpful or even irrelevant with regard to what we might, as philosophers, deem to be more important issues, and also that talking of theories and comics, respectively, as the appropriate units of assessment, is not entirely appropriate – rather we should talk of 'programmes' and 'series', also respectively.[20] However, this further highlights our original concern: what maintains the existence of a given scientific programme as an abstract artefact? Is it the relevant community? Or the last remaining adherent, clinging on to their beliefs as the programme degenerates (think of Priestley and phlogiston, for example)? More generally, how do individual intentions sustain the complex web of interconnections that binds theories into such programmes? With the proliferation of such questions, one might again feel that the view of theories as abstract entities hinders, rather than helps, our attempts to get a grip on the various complexities of scientific practice.

7 Conclusion

Those philosophers of science who take models, and hence – via the Semantic Approach – theories to be abstract, might usefully draw on the view of artworks as abstract artefacts in order to both sharpen and flesh out their account. Nevertheless, there are important differences between science and art that are brought into sharp focus when we begin to think about how this view might mesh with scientific 'discovery' in general and well-known heuristic moves in particular. As we have seen, questions begin to multiply about how such scientific abstract artefacts come to be created and manage to be sustained. It may be, of course, that these questions can be answered, although the ontological cost will almost certainly remain.

The alternative is to chop through this particular Gordian knot by denying the initial assumption, namely that theories are things or entities, abstract or otherwise, to begin with. It is better, I would argue, not to inflate our ontology but instead to pay close attention, nominalistically, to the relevant practices. Here too we may draw on the philosophy of art to help us: just as one can be a kind of eliminativist about musical works (Cameron, 2008), so one can adopt a similar stance towards theories (French and Vickers, 2011). Statements about theories, about their simplicity, their aesthetic qualities, their predictive successes and so forth can still be made, but must be understood as made true by features of the relevant practice (acting as truthmakers) and not by abstract artefacts possessing these qualities. Putting it bluntly, there are no *things* that are theories, but only complex and interwoven practices.

What about the Received View and Semantic Approach that, as sketched at the beginning, characterise theories as deductively closed sets of sentences or families of models respectively? If there are no theories, what are these frameworks characterising? Elsewhere I have argued that these should be regarded as formal devices that we, as philosophers of science, develop and deploy for our own purposes, in order to represent at the meta-level, as it were, the various features of practice at the 'object'-level, that we happen to be interested in (French, 2014). So, if we are keen to talk of 'theories' as true, in the standard Tarskian sense, then we might prefer the sentential or propositional framework; on the other hand, if we want to capture certain kinds of inter-theoretical or theory-data relations, or the relations between mathematics and science, then the Semantic Approach, with its set-theoretic notions of partial isomorphism and homomorphism, seems the better tool for the job. But in neither case are we capturing or representing some entity that is 'out there', in World 3 or as an abstract artefact or whatever. What we are taking as true are certain propositions that are deemed to be significant in some particular scientific context; what we are taking to be inter-related are certain features of the relevant practices. In a sense, one can say that scientists and philosophers of science alike do create or construct theories, not as sustainable abstract artefacts but rather as the nebulous and contestable results of reflections on those practices. Einstein's 'On the Electrodynamics of Moving Bodies', with its talk of clocks and rods, and Minkowski's 'The Fundamental Equations for Electromagnetic Processes in Moving Bodies', with its four-dimensional space-time, are both taken to be formulations or presentations of *the* Special Theory of Relativity. We, both scientists and philosophers of science, may then re-present them in different ways to our chosen audience – students, other scientists, fellow philosophers, whoever – via models, propositions, space-time diagrams, whatever. But there is no thing that the 'the' refers to, no entity that these different formulations and presentations represent. There is only that feature of practice that is Einstein's consideration of clocks and rods, and another that is Minkowski's of space-time, both of which make true certain statements, such as 'Special Relativity is more perspicuously presented via space-time diagrams', or 'Einstein's version is free of realist presuppositions'.

Defending such a stance in detail is a matter for another time; here I simply want to emphasise first that it frees us from being bedevilled by the kinds of questions that have been raised above, and second that the stance itself is the outcome of importing a particular – eliminativist – approach to certain artworks from the philosophy of art into the philosophy of science. Hopefully my discussion has illustrated how relations between philosophy of art and philosophy of science, although problematic in certain respects, may be fruitful in others.

Notes

1 Interestingly (given the discussion to follow), he also takes them to be 'human constructions' whose creation is made possible by the 'symbolic artifacts' of language and mathematics, but which are not to be identified with these artifacts.

2 Gödel is perhaps most famously associated with the espousal of such an intuition, although it is actually not clear what he had in mind (!).

3 There are also interesting things to say about the relationships between Worlds 3 and 2, where there is a kind of 'feedback' effect: 'Our minds are the creators of world 3; but world 3 in its turn not only informs our minds, but largely creates them' (ibid., p. 167). So, for example, the very idea of 'the self' depends on certain views of time that underpin self-identity and exist in World 3.

4 Of course, it could be argued that the sense of surprise here has less to do with discovering some previously undiscovered or unnoticed feature of an entity and more to do with our own limitations with regard to what deductively follows from a given set of premises – that Einstein was surprised by the discovery of $E = mc^2$ should be no surprise, given that lack of deductive transparency.

5 This bears on the issue of creativity and heuristics in music – the view of Mozart in particular as composing through some impulsive, muse-driven creative act over which he had little or no control – which feeds into the 'Eureka' account (if it can even be called such) of creativity – is now seen as a product of nineteenth-century mythologisation of the composer.

6 Of course, when this point is reached may still be debatable. In a famous incident represented in the film *Mr Turner,* by Mike Leigh, when Turner viewed his seascape 'Helvoetsluys' hanging next to Constable's more colourful 'The Opening of Waterloo Bridge', he daubed a red blob in the middle of his sea (Constable's painting featured figures in resplendent red jackets as well as red standards flying) and fashioned it into a buoy, thus completing the painting while it was being exhibited (http://www.gettyimages.co.uk/detail/news-photo/an-employee-looks-at-turners-helvoetsluys-hanging-alongside-news-photo/90996762).

7 This was suggested by a commentator on an earlier incarnation of the material presented here.

8 Likewise, did *Moby-Dick* come into existence only when it was published with that title in New York in November 1851, and not when it was published as *The Whale* in London the month before? Given that the London publisher cut passages, perhaps we *should* take this as a different entity. But Melville himself made last-minute changes – should we take this as being made *to* 'the' work or as each creating a different literary entity?

9 Much discussed in the philosophy of physics in the context of space-time substantivalism; for a useful summary see Norton, 2014.

10 Here Merton is operating at the meta-level in noting that this repeated rediscovery of multiple discoveries in science is caught in a static condition (again, at this level), '…as though it were permanently condemned to repetition without extension' (ibid., p. 238). He then considers the reasons for this resistance to examining the phenomenon by scientists themselves, concluding that these have to do with complex patterns of behaviour associated with, on the one hand, ethical and other issues concerning priority and, on the other, the influence of the 'Eureka' view of scientific discovery. But, of course, this resistance can be, or should be, easily overcome by philosophers, or even sociologists, of science!

11 An anti-realist may indulge in the time-honoured manoeuvre of pointing to cases of underdetermination in response, or, more interestingly, might note that even if we grant that there is a way that the world is, this does not compel its description in the same terms. This again touches on the issue of how we identify the same or different theories and the question whether, in such cases, theories that describe the same unobservable features in different terms are merely nomological variants of one another.

12 Thus, if we take an artwork to be a 'norm kind' in the sense that the artist selects a certain set of properties in creating a work such that those properties become normatively associated with that work (Wolterstoff, 1980), then in the case of literature, a certain fixed sequence of words will determine what counts as a correct instance of the relevant work.

13 This raises the issue of whether translations count as the same work or not. Insofar as they involve different words, in different order, set down by a different person – not. But insofar as they convey the same meaning, refer to the same sequence of events, address the same themes – maybe.

14 There is, of course, much more to say about such differences and how deep they go!

15 Leaving aside issues to do with both men's debts to others such as Barrow (Hall, 2002), a mathematical Platonist will obviously have a ready answer as to how such multiple

discoveries could be possible – indeed, one might wonder why, on this view, there aren't more – while the nominalist, who sees mathematics as no more than a representational device, will presumably point to the similar context of applications for her explanation.

16 Likewise, those who have read his letters and essays would balk at the idea that Tolkien, a Catholic, with his dislike of allegory and his insistence that a mythical world must be constructed via the bedrock of language, could have written a work featuring a Christ-like lion and a mix of mythological creatures from different cultures!

17 According to this theory, there was no 'Big Bang'; rather, the universe undergoes continuous expansion with constant density maintained via spontaneous particle creation.

18 Or tackling the flaw in an old theory, another of Post's list of heuristic moves.

19 Thus some commentators talk about evolutionary theory having itself 'evolved' (yes, very amusing; see: http://evolution.about.com/od/scientists/ss/5-Post-Darwin-Evolution-Scientists.htm). And of course, the very word 'theory' here is loaded in the context of the 'debate' with the anti-evolution zealots!

20 There is comparatively little that has been done in comparing the philosophy of art and the philosophy of science in this regard. Rickles has made a start in this direction by considering whether musical styles might be likened to scientific paradigms (Rickles, 2013). However, as he notes, the driving forces for change are, at the very least, different, and we don't seem to have anything equivalent to a crisis or anomaly in a musical style.

References

Adlam, E. (2011). 'Poincaré and Special Relativity'. Available at: http://arxiv.org/pdf/1112.3175v1.pdf.

Bowler, P. J. (1976). 'Alfred Russell Wallace's Concepts of Variation'. *Journal for the History of Medicine,* 31, 17–29.

Bulmer, M. (2005). 'The Theory of Natural Selection of Alfred Russel Wallace FRS'. *Notes and Records of the Royal Society,* 59, 125–36.

Cameron, R. (2008). 'There Are No Things That Are Musical Works'. *British Journal of Aesthetics,* 48, 295–314.

Castelvecchi, D. (2014). 'Einstein's Lost Theory Uncovered'. *Nature,* 506, 418–19.

da Costa, N.C.A. and French, S. (2003). *Science and Partial Truth: A Unitary Approach to Models and Scientific Reasoning.* New York: Oxford University Press.

Eckert, M. (2014). 'How Sommerfeld Extended Bohr's Model of the Atom (1913–1916)'. *The European Physical Journal,* H (39), 141–56.

Frankel, H. (1987). 'The Continental Drift Debate', in H.T. Engelhardt Jr and A.L. Caplan. (eds.) *Scientific Controversies: Case Solutions in the Resolution and Closure of Disputes in Science and Technology.* Cambridge: Cambridge University Press.

French, S. (2014). *The Structure of the World: Metaphysics and Representation.* Oxford: Oxford University Press.

French, S. and Vickers, P. (2011). 'Are There No Things That Are Scientific Theories?' *British Journal for the Philosophy of Science,* 62, 771–804.

Gay, H. (1978). 'The Asymmetric Carbon Atom: (a) A Case Study of Independent Discovery; (b) An Inductivist Model for Scientific Method'. *Studies in History and Philosophy of Science,* 9, 207–38.

Giere, R. (1988). *Explaining Science: A Cognitive Approach.* Chicago, IL: University of Chicago Press.

Giere, R. (2008). 'Why Scientific Models Should Not Be Regarded as Works of Fiction', in M. Suarez (*ed.*) *Fictions in Science: Philosophical Essays on Modelling and Idealization.* London: Taylor and Francis.

Hall, A. R. (2002). 'Newton Versus Leibniz: From Geometry to Metaphysics', in I.B. Cohen and G.E. Smith (eds.) *The Cambridge Companion to Newton*. Cambridge: Cambridge University Press, 431–54.

Katzir, S. (2005). 'Poincaré's Relativistic Physics: Its Origins and Nature'. *Physics in Perspective,* 7, 268–92.

Lutz, S. (2014). 'What's Right with a Syntactic Approach to Theories and Models?' *Erkenntnis,* 79, 1–18.

Merton, R.K. (1963). 'Resistance to the Systematic Study of Multiple Discoveries in Science'. *European Journal of Sociology,* 4 , 237–82.

Meskin, A. (2012). 'The Ontology of Comics', in A. Meskin and R.T. Cook (eds.) *The Art of Comics: A Philosophical Approach*. London: Wiley, 31–46.

Nicholson, A.J. (1960). 'The Role of Population Dynamics in Natural Selection', in S. Tax (ed.) *Evolution after Darwin*, Volume 1. Chicago, IL: Chicago University Press, 477–522.

Norton, J.D. (1984). 'How Einstein Found His Field Equations: 1912–1915'. *Historical Studies in the Physical Sciences*, 14, 253–316; reprinted in D. Howard and J. Stachel (eds.) (1989). *Einstein and the History of General Relativity: Einstein Studies*, Volume 1. Boston: Birkhäuser, 101–59.

Norton, J.D. (2014). 'The Hole Argument', in Edward N. Zalta (ed.) *The Stanford Encyclopedia of Philosophy*. (Spring 2014 Edition). Available at: http://plato.stanford.edu/archives/spr2014/entries/spacetime-holearg/.

Popper, K. (1972). *Objective Knowledge: An Evolutionary Approach*. Oxford: Clarendon Press.

Popper, K. (1978). Three Worlds: The Tanner Lectures on Human Values. Utah: University of Utah Press.

Post, H.R. (1971 [1993]). 'Correspondence, Invariance and Heuristics'. *Studies in History and Philosophy of Science*, 2, 213–55. Reprinted in French, S. and Kamminga, H. (eds.) (1993). *Correspondence, Invariance and Heuristics: Essays in Honour Of Heinz Post*. Boston Studies in the Philosophy of Science, Volume 148, Dordrecht: Kluwer Academic Press, 1–44.

Rickles, D. (2013). 'Musicology as an Object for HPS? An Exploration'. *Aesthetics On-Line*: http://aesthetics-online.org/?page=RicklesMusicology&hhSearchTerms=%22Rickles%22.

Rorhbaugh. G. (2005). 'I Could Have Done That'. *British Journal of Aesthetics,* 45, 209–28.

Schindler, S. (2008). 'Model, Theory, and Evidence in the Discovery of the DNA Structure'. *The British Journal for the Philosophy of Science*, 59, 619–58.

Suppe, F. (ed.) (1977). *The Structure of Scientific Theories*. Urbana, IL: University of Illinois Press.

Suppe, F. (1989). *The Semantic View of Theories and Scientific Realism*. Urbana and Chicago, IL: University of Illinois Press.

Thomasson, A. (1999). *Fiction and Metaphysics*. Cambridge: Cambridge University Press.

Thomasson, A. (2004). 'The Ontology of Art', in P. Kivy (ed.) *The Blackwell Guide to Aesthetics*. Oxford: Blackwell.

van Wyhe, J. (2013). *Dispelling the Darkness: Voyage in the Malay Archipelago and the Discovery of Evolution by Wallace and Darwin*, Singapore: World Scientific.

Wolterstoff, N. (1980). *Works and Worlds of Art*. Oxford: Clarendon Press.

Zahar, E. (2001). *Poincaré's Philosophy*. Chicago, IL: Carus Publishing Company.

Index